The Shortwave Listener's Q & A Book

The Shortwave Listener's Q & A Book

Anita Louise McCormick

TAB Books
Division of McGraw-Hill, Inc.
New York San Francisco Washington, D.C. Auckland Bogotá
Caracas Lisbon London Madrid Mexico City Milan
Montreal New Delhi San Juan Singapore
Sydney Tokyo Toronto

Trademarks

Bearcat	Uniden Corporation
Drake	R. L. Drake Co.
Icom	Icom Inc., Osaka, Japan
Microdec	Somerset Electronics, Inc.
Panasonic	Matsushita Electric Corporation of America
Radio Shack	Tandy Corporation
Sangean	Sangean America, Inc.
World Radio TV Handbook	Billboard Books

FIRST EDITION
FIRST PRINTING

Library of Congress Cataloging-in-Publication Data

McCormick, Anita Louise.
The shortwave listener's Q & A book / by Anita Louise McCormick.
p. cm.
Includes index.
ISBN 0-07-044774-8 (p)
1. Shortwave radio—Amateurs' manuals. I. Title. II. Title: Shortwave listener's Q and A book.
TK9956.M377 1994
384.54—dc20 93-48717
CIP

Acquisitions editor: Roland S. Phelps
Editorial team: Andrew Yoder, Managing Editor
John T. Arthur, Editor
Production team: Katherine Brown, Director
Jan Fisher, Desktop Operator
Sandy Fisher, Proofreader
Joann Woy, Index
Stephanie Myers, Computer Illustrator
Design team: Jaclyn J. Boone, Designer
Brian Allison, Associate Designer

EL1
0447748

To all the wonderful people who
make shortwave listening such an enjoyable hobby.

Contents

Acknowledgments

Thanks to all the people, organizations, and companies that helped, with materials or encouragement, to make this book possible, including the A*C*E, the ARRL, Bearcat Radio Club, R. L. Drake Company, Grove Enterprises, Grundig, Chuck Harder, Icom, Lowe Electronics, MFJ Enterprises, Inc., Monitoring Times, NASA, NASWA, Offshore Echo's, Popular Communications, Radio Caroline, Radio For Peace International, Radio New York International, Panasonic, Radio Shack, Sangean, Universal Radio, Inc., Voyager Broadcast Services, Andrew Yoder, and, of course, Allan H. Weiner.

Foreword

In times of civil strife and war, the mass media outlets almost always announce that one of their sources of information was the "national radio" from whatever country is involved. Actually, much of this information is gleaned from shortwave radio stations that can be heard around the world. In fact, at least one of the U.S. television networks had a monitoring station set up during the Persian Gulf Conflict, just to acquire information from these shortwave programs. In spite of the wealth of news that is available free-of-charge to the public on the shortwave bands, few people in North America take advantage of this resource.

Because signals can be heard around the world on shortwave, you can hear programs from many different countries, in about 100 different languages, and from numerous cultural and political perspectives. Other than actually traveling to each country and spending time there, the next best way to learn about these countries is to listen to shortwave radio. Even elementary and high school teachers have noticed the benefits of using shortwave radio in the classroom to teach social studies.

In spite of these advantages, shortwave has gained little popularity. Radio stations have been broadcasting on shortwave since the 1920s, yet the subject is still mysterious and confusing to the general public. Perhaps the problem is a lack of publicity for shortwave or the lack of shortwave radios in the marketplace. Or perhaps shortwave radios are priced too high to attract the casual listener.

Even if these problems have hampered the popularity of shortwave radio in the past, times are changing. Modern shortwave receivers are small, light, and uncomplicated, and the prices are dropping fast. The Persian Gulf Conflict and the more-recent antics of some talk show hosts have placed shortwave radio in the spotlight.

That leaves the general public with an affordable "new" technology . . . and a great many questions. In order to quench this thirst for information, long-time shortwave radio enthusiast, Anita McCormick, wrote this book in a "question and answer" format. This book is perfect for the novice who wants to learn more about what can be heard on the shortwave bands, but doesn't want to wade through several books to get a grasp on the subject.

I greet this book with the great anticipation that it will enlighten others about the world we live in. I hope that it will spark interest in the shortwave listening hobby and that thousands more will experience the current offering of worldwide radio programming. Shortwave radio listening is a great educational tool and source of entertainment that has been underutilized for years.

Andrew Yoder

Introduction

These days, almost everyone has heard of shortwave radio. It's a magical medium that makes it possible to hear all those exotic locations overseas. If you own a shortwave radio, The British Broadcasting Corporation, Radio Moscow, Radio Japan, and dozens of other stations are as accessible to you as your local AM and FM broadcasters.

Though "shortwave" is getting to be a household word here in the United States, many people have questions about exactly what shortwave radio is, what you can hear on shortwave, and what it costs to get into the hobby.

This book will answer the most commonly asked questions about shortwave listening, and show you how easy and inexpensive it is to obtain a shortwave radio and hear programs from all over the world, many in English.

Chapter 1 gives you a brief introduction to shortwave radio listening. You'll find out how easy it is to be a shortwave listener and get in on the many exciting benefits that millions of people around the world are already enjoying.

Chapter 2 tells you about the stations broadcasting on shortwave, who runs them, why they're on shortwave, and the kinds of programs they broadcast. You'll learn about government-operated stations, privately run commercial stations, religious stations, special events stations, offshore broadcast stations, even pirate stations!

Chapter 3 lets you in on the secrets of communicating with international shortwave stations. You'll learn how to write letters that get read on the air, write letters that influence programming decisions, and how you can build a nice collection of cards, pennants, and other souvenirs from the stations you hear.

Chapter 4 explores the world of ham radio and utility stations, which are on the air to conduct 2-way communication between individuals, companies, and governments. You'll find out how to tune in boats, planes, press services, and ham radio operators. You'll also learn about decoders, which can be used to translate the "dits" and "dahs" you hear on shortwave into readable messages.

Chapter 5 explains in non-technical language how radio waves make the trip from one place to another. You'll read about shortwave "skip," shortwave relay stations, propagation reports, and the effect of the sun on shortwave radio reception.

Chapter 6 introduces you to the various shortwave radios available on the market, and helps you to decide which one is right for your budget and listening interests. You'll find out what to look for in a receiver, and the approximate price you can expect to pay. This chapter also describes some inexpensive, easy-to-erect antennas that will let you hear even MORE fascinating broadcasts from around the world.

Chapter 7 shows you how to keep up with the latest information on shortwave listening. You'll read about the magazines, radio programs, and clubs that dedicated shortwave listeners use to keep current with the world of international broadcasting.

This book also contains appendices that make it easy to look up station addresses, shortwave mail-order dealers, and radio clubs. It also includes a list of popular books that will help you get more out of your shortwave listening hobby.

Happy listening!

1
What is shortwave radio?

Q. I can hear plenty of music, news, and talk shows on the local radio stations in my area. What's so special about shortwave radio?

A. AM and FM radio stations are fine for providing music and news about your region. But a shortwave radio (Fig. 1-1) gives you the WORLD! You can switch on a shortwave radio, turn a few knobs or push a few buttons and hear programs from England, Russia, Israel, Brazil, . . . almost anywhere on Earth!

Dozens of international broadcast stations can be heard day and night on the shortwave bands, and most transmit at least some of their programming in English. Even the cheapest shortwave radios on the market will bring in a good number of foreign stations, and can open an exciting new world of news, entertainment, and information to you.

Q. Do you need special equipment to hear foreign broadcasts?

A. You'll need to buy a shortwave radio. They can be found at many electronics and department stores. Radio Shack stores across the USA carry a nice variety of shortwave radios and accessories.

If you can't locate a receiver you'd like to own, check with some of the mail-order companies that specialize in selling shortwave radios and related products. A few of the larger and better-known companies in this field are EEB (Electronics Equipment Bank), Grove Enterprises (Fig. 1-2), and Universal Radio. Nearly all shortwave mail-order dealers offer free catalogs that feature photos and full descriptions of all the shortwave radios and related products they sell. Mail-order companies that cater to the international listener are listed in Appendix C.

Q. Are shortwave radios expensive?

Grundig (Lextronix, Inc.)

1-1 This Grundig YB 235 can bring in shortwave stations from around the world.

Grove Enterprises, Inc.

1-2 Bob Grove, President of Grove Enterprises, Inc., enjoying his shortwave listening hobby.

A. They certainly don't have to be! Although you can spend hundreds, even thousands, of dollars on shortwave equipment, it is possible to get into shortwave listening for under $100. In the past few years, several manufacturers have introduced low-priced pocket portables that can give you a taste of international shortwave listening for around $30 to $50 (Fig. 1-3). Although these low-priced models don't have all the fancy features of their more expensive counterparts, they can provide you with endless hours of listening enjoyment.

1-3 Low-priced portable receivers put shortwave listening within everyone's budget.

Q. Don't the national TV and radio networks and newspapers tell me everything I need to know about events overseas?

A. Only if you're satisfied with a 5- or 10-minute rundown of what has happened on our entire planet in the past 24 hours! With shortwave radio, you will have the advantage of hearing first-hand information about dozens of foreign countries. You'll find out how other countries see themselves, their problems, and their position in the world . . . and you'll find out what they think about our country, as well. You'll hear background reports on international events and learn the "how and why" of news stories that the American press barely mentions.

With a shortwave radio, you can be your own news director. Instead of letting the major American broadcast networks decide what is important to you, you can decide for yourself! If you want to know what's going on with the Russian economy, tune in Radio Moscow. If you're interested in hearing more about a scandal within the Royal family, tune in the British Broadcasting Corporation (BBC). And if you want to find out more about the newly elected Israeli government, tune in KOL Israel.

Q. What can I hear on shortwave?

A. Most everything! You can hear local and world news, foreign press reviews, music, talk shows, sports, contests, "mailbag" programs of listener's letters, travel programs, business news, environmental updates, religious broadcasts, ham radio operators, unlicensed "pirate" stations, ship-to-shore transmissions, you name it! With a shortwave radio, you'll have more entertainment options than you ever dreamed of.

Q. Do shortwave stations have promotional items to send out to their listeners?

A. Yes! As a shortwave listener, you can collect great souvenirs from the stations you hear. Stickers (Fig. 1-4), postcards, pennants, and calendars are just a few of the items you can request from your favorite shortwave stations.

1-4 Many shortwave stations offer colorful stickers to their listeners.

In addition to that, a number of shortwave stations and programs, including Radio New York International (Fig. 1-5), have T-shirts available for their listeners to purchase. You can also buy great shortwave-related items, such as the Universal Shortwave "Listen to the World" mug in Fig. 1-6, from many shortwave mail-order dealers.

Q. Are many programs broadcast in English?

A. Nearly all the major international shortwave stations broadcast at least some of their programs in English, so you don't have to be multi-lingual to understand what they're saying.

If you ARE interested in learning a new language, several shortwave stations offer free over-the-air language courses. And, as a foreign language student, you'll be

1-5 This Radio New York International T-shirt features a drawing of their former broadcast ship, *The M/V Sarah.*

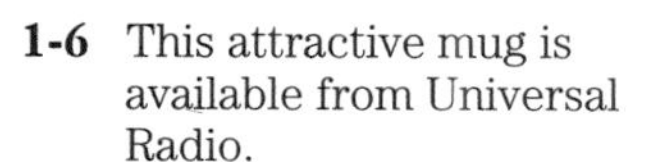

1-6 This attractive mug is available from Universal Radio.

able to find many native speakers on the air—especially if you're learning Spanish, French, German, Japanese or Chinese.

Q. How can I find the country I want to hear?

A. There are many sources of schedule information available to shortwave listeners. *Passport to World Band Radio* and *The World Radio TV Handbook* are published on an annual basis. They're filled with information on the times and frequencies (number on the dial) of shortwave radio broadcasts. *Monitoring Times* and *Popular Communications* magazines publish monthly schedules of shortwave broadcasts in English. As an additional source of information, you might want to join a shortwave listener's club. Most clubs publish a monthly bulletin, where members can exchange information about schedule changes, new stations and shortwave listening in general.

Table 1-1 is a list of frequencies major shortwave stations are currently using for English broadcasts to North America.

Table 1-1. Shortwave English language broadcast frequencies

Station/country	Frequency in MHz
Albania	9.580, 11.840
RAE, Argentina	15.345
Australia	11.855, 13.755, 17.750, 17.840, 21.525
ORF/Austria	6.015, 9.870, 9.880
BBC/England	5.975, 6.175, 6.195, 7.325, 9.515, 9.915, 11.965, 12.095, 15.260
R. Sofia/Bulgaria	11.720, 15.330, 17.825
RCI/Canada	5.960, 6.120, 9.535, 9.755, 11.845, 11.940, 11.955, 13.670, 17.820
China	9.870, 11.715, 11.840
RFPI, Costa Rica	7.375, 7.385, 13.630, 15.030
R. Habana/Cuba	6.010, 6.180, 9.655, 9.815, 13.660
Czech Republic	5.390, 7.345, 9.485, 9.810, 11.990, 13.715, 17.535
HCJB/Ecuador	9.745, 11.925, 15.115, 17.890, 21.455
R. Cairo, Egypt	9.475, 11.660
Finland	11.755, 15.185, 15.400, 21.550
Deutsche Welle, Germany	6.040, 6.085, 6.145, 7.285, 9.615, 9.690, 9.700
Guatemala	3.300
Greece	9.380, 9.420, 11.645
R. Netherlands/Holland	6.020, 6.165, 11.835
R. Budapest, Hungary	5.970, 9.835, 11.910, 15.220
R. Baghdad, Iraq	11.810, 15.180, 17.940
KOL Israel	7.465, 9.435, 11.587, 11.603, 15.605, 15.640, 17.575, 17.590
RAI, Italy	9.575, 11.800
R. Japan	5.960, 11.815, 11.840, 11.860, 15.195, 17.775
Norway	9.675, 15.165
Romania	6.155, 9.510, 9.570, 11.830, 11.940
R. Moscow/Russia	7.205, 7.335, 9.505, 9.530, 9.625, 9.675, 9.860, 9.905, 11.790, 15.280
New Zealand	9.700, 11.735, 15.120, 17.770
Nigeria	7.225

North Korea	11.335, 13.760, 15.130
Norway	9.560, 9.675, 15.165
S. Korea	7.550, 9.570, 13.670, 15.575
Spanish National Radio	9.525
R. Sweden	9.695, 11.820, 15.190, 15.240, 21.500
Swiss Radio Intl.	6.135, 9.650, 9.855, 12.035, 17.730
VOFC, Taiwan	5.950, 9.680, 11.740, 15.345
Turkey	9.445
R. Kiev, Ukraine	9.685, 11.720, 15.180, 15.195, 15.580
Christian Science/USA	5.850, 9.350, 9.455, 13.760, 15.665
KTBN/USA	7.510, 15.590
KVOH/USA	9.785, 17.775
WHRI/USA	7.315, 9.495
WINB/USA	15.185, 15.295
WMLK/USA	9.465
WRNO/USA	7.335, 15.420
WWCR/USA	5.810, 5.935, 7.435, 13.845, 15.685
WYFR/USA	5.950, 5.985, 7.355, 9.705, 11.830, 15.335, 17.760, 21.525, 21.615
Voice of America	5.995, 6.130, 7.405, 9.455, 9.775, 11.580, 15.120, 15.205, 17.735, 21.550
Vatican Radio	7.125, 9.600, 9.650, 11.830

Q. How are shortwave broadcasts heard so far from the country of their origin?

A. Radio waves in the shortwave band are capable of covering much greater distances than signals broadcast on the AM and FM bands. They are reflected off electrically charged layers of the atmosphere, (known as the ionosphere), 60 to 200 miles above us (Fig. 1-7). When radio waves bounce off the ionosphere and come back to Earth, they can be heard by listeners hundreds, if not thousands, of miles away from the transmitter site.

To make their broadcasts even clearer, some stations make use of satellite technology and earth-based relay stations to deliver their programs to shortwave listeners around the world (Fig. 1-8).

Q. Do I need to put up an antenna to hear foreign stations?

A. The extendable, built-in "whip" antenna that comes with all pocket portables will let you hear dozens of shortwave stations. As a new listener, that's all you'll really need. After you're gained some experience with international listening and want to tune in the more exotic, hard-to-hear stations, you'll probably want to install an outdoor antenna. Radio Shack stores nationwide sell several types of ready-made shortwave antennas. An even larger selection of external shortwave antennas is available from the mail-order companies listed in Appendix C.

Q. What is a "short wave" anyway?

A. The shortwave band is between the AM and FM bands on the electromagnetic frequency spectrum. It is used for international broadcasts because the radio waves

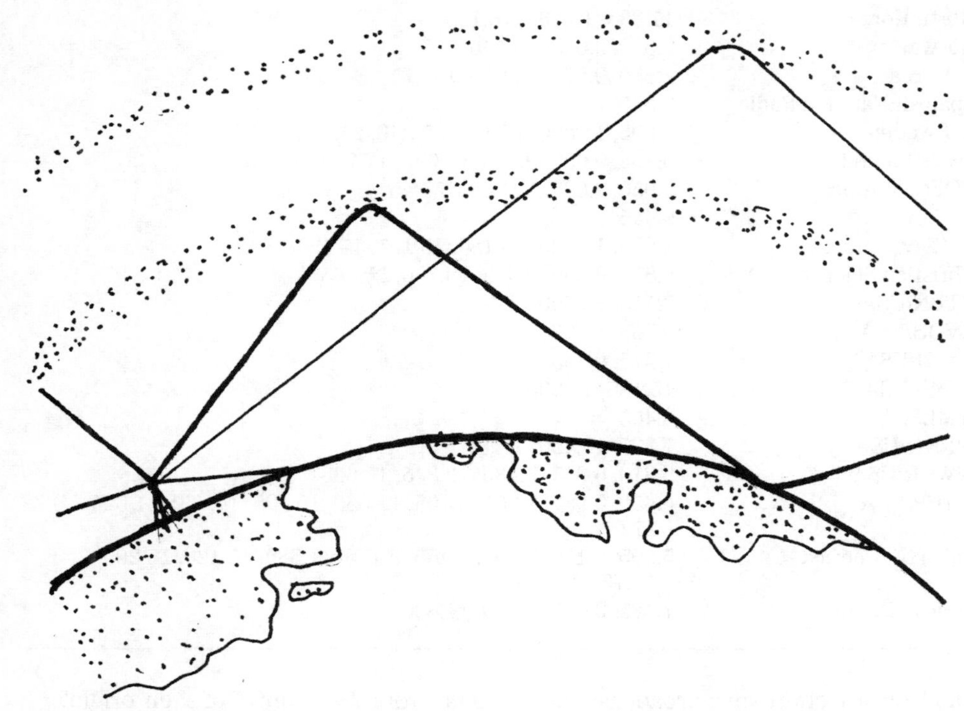

1-7 Once a shortwave signal hits the ionosphere, it can "skip" over great distances.

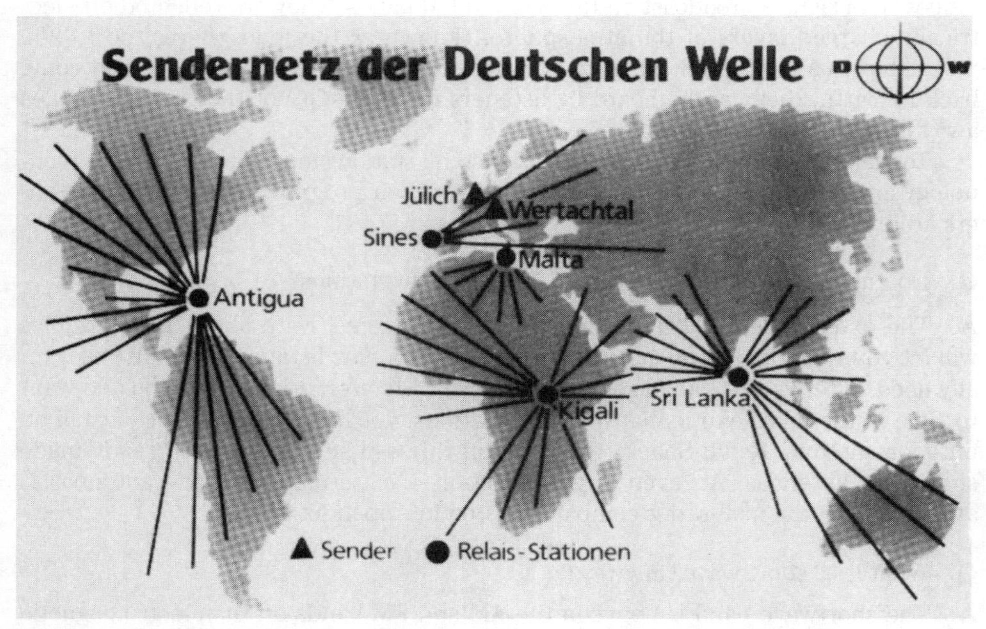

1-8 Deutsche Welle, The Voice of Germany, operates relay stations in many parts of the world.

transmitted on these frequencies "skip" over long distances. The shortwave band is generally considered to be between 3 and 30 MHz. If you've ever used a CB, which operates in the 27 MHz band, you were operating on a shortwave frequency.

Q. How long have people been broadcasting on shortwave?

A. Like many breakthroughs in the field of radio communications, much of the early knowledge of the shortwave frequency bands came from experiments conducted by amateur radio operators. They tried broadcasting on various parts of the radio spectrum in order to see which frequencies made it possible to communicate over the greatest distances.

Once the usefulness of shortwave was established, both government and private stations used shortwave radio to broadcast to people all over the world. By the time World War II began, many governments were using shortwave radio to transmit their views to a global audience (Fig. 1-9). People who had never before been interested in shortwave bought receivers so they could hear up-to-the-minute reports from the battlefields overseas.

1-9 A World War II radio map of Europe.

In the years that followed, more and more shortwave stations were put on the air to serve a wide variety of purposes. Many governments wanted a shortwave voice to the world, so they set up powerful transmitters and huge antenna towers so the largest possible audience could hear them. Religious organizations saw the value of shortwave broadcasting as a powerful tool for getting their message out. And commercial shortwave stations, hoping to make a profit by selling blocks of air time, took to the international airwaves as well.

During the Gulf Conflict, shortwave radio listening experienced a new resurgence of popularity—especially here in the USA. Listeners around the world tuned in news of the latest developments and conflicts on Mid-east stations such as KOL Israel (Fig. 1-10), Radio Kuwait (Fig. 1-11), Radio Cairo, and Radio Iraq International.

Q. Do you have to be electronically inclined to get into shortwave listening?

A. Not at all! You don't have to know what makes televisions, telephones, or computers work in order to enjoy them. The same goes for shortwave radios! Many low

KOL ISRAEL
External Service
P.O.B. 1082
91 010 Jerusalem, Israel

THANK YOU FOR YOUR RECEPTION REPORT
NOUS VOUS REMERCIONS POUR VOTRE RAPPORT D'ÉCOUTE
AGRADECEMOS SU INFORME DE RECEPCIÓN

11588 kHz Date/día 7.12.91 10500 GMT

YOUR REPORT HAS BEEN CHECKED AND AGREES WITH OUR LOG
VOTRE RAPPORT A ETE VERIFIE ET TROUVE CONFORME A NOS SPECIFICATIONS
SU INFORME HA SIDO VERIFICADO Y CONCUERDA CON NUESTROS DETALLES

1-10 A card from KOL Israel.

1-11 A card from Radio Kuwait.

and medium-priced portables (Fig. 1-12) are very easy to operate. You turn them on, turn the volume up and tune until you come upon a station that you want to hear. That's all there is to it.

Q. Is it really that simple?

A. Sure! Anyone can operate a shortwave radio—especially the less expensive pocket-sized models. Shortwave radio listening is a fun and exciting hobby for people of all ages. Everyone can enjoy it.

Q. If it's that easy, why don't more people listen to shortwave?

A. Probably because they don't know how easy it is . . . and how much fun it can be to tune in programs from all over the globe. But the good news is that shortwave listening is on the upswing in the United States. A recent survey taken by the British Broadcasting Corporation found that about 17 million shortwave radio sets are in use in the USA. This compares with 20 million in the United Kingdom, 300 million in the former Soviet Union, and 1 billion worldwide.

So, if your friends don't know how great shortwave listening is, do them a favor and tell them!

Sangean America, Inc.

1-12 This Sangean portable is a nice starter radio.

2
Shortwave stations

Q. What kinds of stations broadcast on the shortwave bands?

A. Most of what you'll hear on shortwave will fall into one of the following categories:

- Government-operated stations
- Privately operated commercial stations
- Religious broadcasters
- Pirate stations
- Offshore stations
- Special events stations

Q. What kind of programming do government-operated shortwave stations broadcast?

A. Government-operated stations air their national views to shortwave listeners around the world. The Voice of America (Fig. 2-1) is the United States government's shortwave outlet. They have been on the air for over 60 years.

During the cold war, Radio Moscow (Fig. 2-2), as well as other hard-line communist stations, such as Radio Berlin International, Radio Prague, and Radio Budapest International (Fig. 2-3), used shortwave broadcasting to try to convince listeners around the world that their political system was best. In order to keep listeners, they mixed news and lectures on political ideology in with programs of music, science, culture, travel features, contests, and "Mailbag" shows where listener's letters were read and questions answered.

Since the cold war ended, shortwave broadcasting in general has become more friendly and informal. With fewer political differences to argue, government-owned stations can make more time for lively and informative programs on such topics as

2-1 A card from The Voice of America.

popular music, local festivals, new products, astronomy, ecology, sports, and international business news.

All of the former Soviet republics are doing their best to use shortwave radio as a means of attracting business investment to their country. Radio Ukraine International, Radio Vilnius—Lithuania, and a few other formerly Soviet countries are easy to hear in North America.

News broadcasts on government-owned stations are not as one-sided as you might imagine. In addition to their basic national and international news broadcasts, most government-owned shortwave stations air programs of editorial opinion and press reviews that examine problems and issues from many different viewpoints. Deutsche Welle, The Voice of Germany (Fig. 2-4), for example, devotes much time to presenting programs such as *European Journal*, *Talking Point*, and *Commentary* that examine the issues facing Germany, Europe, and the world.

Developing nations in many regions are making increasing use of shortwave to promote their countries and boost their economies. The Voice of Nigeria (Fig. 2-5), one of the most powerful stations broadcasting out of Africa, can easily be heard in North America late at night.

2-2 A card from Radio Moscow.

Budget cutbacks have forced some government-owned shortwave stations to curtail special programming for overseas listeners. Instead, they air domestic programs (or Home Service) over their shortwave transmitters. This means you hear the same thing as the "natives," which can be pretty interesting if you know the language!

Radio New Zealand International (Fig. 2-6) has been airing their domestic "National Radio" service on shortwave for a few years now. And since the dominant language in New Zealand is English, they have quite a few American listeners. National Radio's program lineup includes news, weather forecasts, music for all tastes, talk shows, gardening tips, rugby games, and special programs for listeners on nearby Pacific islands. They also broadcast programs for the nation's native Maori population, and send New Zealand's children off to school with a smile on their faces with their Monday through Friday broadcasts of "Ears" and "School Bus Story."

Q. What can I hear on privately-owned commercial shortwave stations?

A. It depends on which one you're listening to! Privately-owned commercial shortwave stations basically fall into two categories: stations that re-broadcast an AM or FM service to reach a larger audience, and stations that sell blocks of time to make a profit.

2-3 A card from Radio Budapest.

CFRX Toronto, Ontario, Canada (Fig. 2-7) relays the programs of Toronto AM station CFRB-1010 twenty-four hours a day over their 1,000 watt transmitter on 6070 KHz. This makes it possible for shortwave listeners to hear them throughout much of the USA and Canada. Their program lineup includes news, talk shows, and some light music.

Several commercial shortwave stations in the USA sell blocks of time to religious organizations and other people who would like to be on the international airwaves, but can't afford to have a station of their own. WWCR Nashville, TN, WHRI South Bend, IN, WRNO New Orleans, LA, KCBI Denton, TX and KVOH Los Angeles, CA (Fig. 2-8) all operate this way. They can easily be heard throughout North America, as well as in most other parts of the world.

Several national American talk shows are now buying time on commercial shortwave stations. Chuck Harder's (Fig. 2-9) "For The People" program can be heard over WHRI between 2–4 PM on 9.485 MHz and between 10–12 PM on 7.315 MHz Eastern time. People from all over call in to discuss government and consumer issues. "For the People" also publishes a bi-weekly paper, *The News Reporter*, which you can order through their organization.

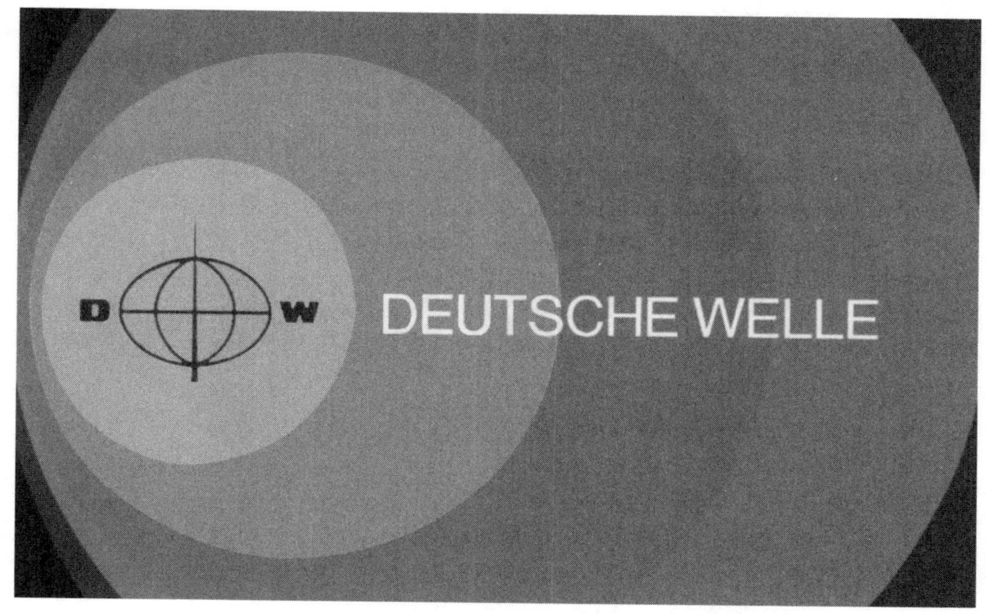

2-4 A card from Deutsche Welle, The Voice of Germany.

2-5 A card from The Voice of Nigeria.

Radio New Zealand International

Thank you for your report on our transmission

on 5 Oct. 1991 on a frequency of 9700

kHz which we are pleased to verify.

Our 100 kw transmitter is located

at Rangataiki near Taupo.

'Calling Japan' pilot series

tiki card

GREETINGS FROM NEW ZEALAND

P.3060

2-6 A card from Radio New Zealand International.

2-7 A card from CFRB/CFRX Toronto, Canada.

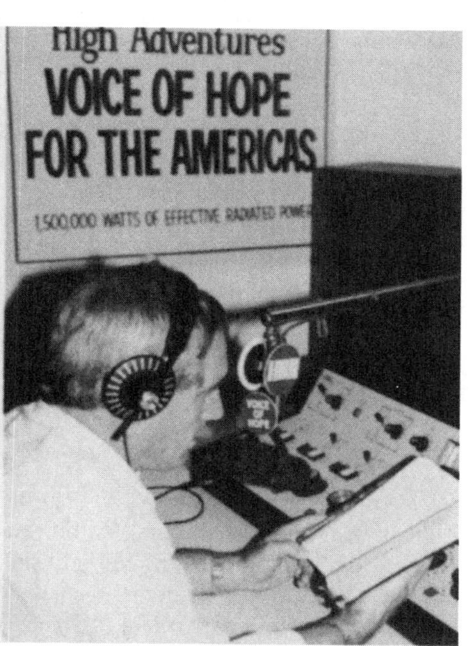

2-8 A card from KVOH—The Voice of Hope, Los Angeles, CA.

For the People

2-9 Chuck Harder's "For the People" program is heard over WHRI—Noblesville, IN.

Q. What kind of programs do non-commercial religious stations broadcast?

A. As opposed to commercial stations that sell time to religious organizations to make a profit, non-commercial religious stations operate primarily to "get the message out." They do this through presenting a wide variety of programs, both religious and general interest, to their listeners.

HCJB Quito, Ecuador (Fig. 2-10), is one of the most popular stations in this category. In addition to many fine religious programs, HCJB's schedule includes programs of news of South America and the world, Equadorian music, mailbag shows, shortwave listening tips, nature and science programs, a cooking show, and programs about Equadorian history and culture.

Q. What are pirate stations?

A. Pirate stations are low-power unlicensed radio stations, operated by hobbyists that have an irresistible urge to broadcast (Fig. 2-11). Their programs usually consist of talk, rock music and comedy skits—sometimes written and performed by the station operators themselves.

Shortwave pirate stations can usually be found during the evening on weekends or holidays between 7.415 and 7.465 MHz. Since their broadcasting activities are illegal, and getting caught by the Federal Communications Commission (FCC) could mean several thousand dollars in fines, confiscation of their equipment, and revocation of any

2-10 A card from HCJB—Quito, Ecuador.

FCC licenses they possess, operators tend to keep their programs short and infrequent. Most pirate programs last only 15 to 30 minutes and, for security reasons, broadcasts are usually not announced to the shortwave community ahead of time. So if you want to hear a pirate, you'll have to stay tuned to their frequencies and be patient!

2-11 Pirates like to entertain their listeners.

Pirates receive mail from their listeners by way of a maildrop, a trusted friend (usually someone that lives in an entirely different part of the country) that volunteers to receive their mail and forward it on to them. It's perfectly legal to write to pirate stations, and if you enclose return postage, they'll often be happy to send some souvenir items for your collection (Fig. 2-12).

2-12 Pirate station souvenirs.

Some pirates air programs recorded by like-minded hobbyists who either don't have a transmitter of their own or are not willing to take the risk of getting caught. The Canadian pirate station CSIC (pronounced sea-sick) airs programs produced by Radio Scottish Montreal (Fig. 2-13) and Radio North Coast International. They can often be heard well throughout much of North America.

Redio Montreal Albanach
Radio Scottish Montreal

La voix de la communauté Ecossais à
Montréal est Redio Montreal Albanach.

Suas le Quebec Albanach!

Congratulations! You have heard Redio Montreal Albanach/Radio Scottish Montreal and we can confirm your reception as follows:

Frequency 7413 Khz. (via CSIC Transmitter)

Time 0203-0234 UTC Date: 21 June 1992

QSL Number 11 Rob Roy

Radio Scottish Montreal- news. music and information about Montreal's Scottish community.

2-13 A card from Radio Scottish Montreal.

There are enough pirate radio enthusiasts in the USA and Canada to support a club, The Association of Clandestine Enthusiasts (The A*C*E), and two annual books: *The Pirate Radio Directory* (Fig. 2-14), by George Zeller and Andrew Yoder and *The Worldwide Pirate Radio Logbook*, by Andrew Yoder. In Europe, pirate radio has an even larger following.

Q. What are offshore stations?

A. Exactly what they sound like, they're floating radio stations built into ships!

In the late 1950s when most European governments only permitted government-controlled stations to broadcast, several groups of radio enthusiasts constructed offshore stations and anchored them in international waters near England, Denmark, and a few other countries. They sold commercial time to finance their ventures and played the latest rock music from Europe and North America on AM, FM, and shortwave, giving their European audience a real listening alternative. Offshore radio forced European stations to modernize and play the "new sounds" in rock music that listeners demanded.

2-14 *The Pirate Radio Directory.*

Radio Caroline (Fig. 2-15), one of the longest-lived offshore stations, started broadcasting in 1964. Their DJs played the Beatles, the Rolling Stones, and other popular groups that were not aired on government-run stations. They continued to broadcast until the late 1980s, when they were forced off the air by the combined efforts of several government agencies. The Dutch government recently returned Radio Caroline's transmitting equipment and the Caroline crew are exploring ways of going back on the air without encountering legal problems. They are financing their efforts by asking for donations from former listeners and offering station-related merchandise through their Horizon Sales catalog (Fig. 2-16).

A new kind of offshore radio station came close to going on the air in early 1994. The *M/V Fury*, the broadcast ship of Voyager Broadcast Services (Fig. 2-17), was licensed and registered with the Central American nation of Belize. V.B.S. planned to sell blocks of airtime to religious organizations, as well as other people and groups that wanted access to the airwaves. They built the station on a ship because the unique signal-reflecting properties of the water would give the transmitters more range. Unfortunately the FCC, claiming that the *Fury's* transmitters had radiated some stray signals while they were being tested, disassembled and confiscated the transmitters only days before the ship was to leave its South Carolina port for Belize.

Allan Weiner (Fig. 2-18), an engineer with V.B.S., had hoped to air his popular Radio Newyork International (R.N.I.) broadcasts over this station. RNI had broadcast from international waters near New York City for four days in 1987 until the FCC and other government agents moved in and took over the ship, forcing the station off the air. RNI is exploring other ways of returning to the air, so stay tuned.

Radio Caroline

2-15 A Radio Caroline reception verification card and bumper sticker.

Q. What is a special events station?

A. A station that operates only on a day (or days) that commemorate a special event. One of the best known special events stations is Radio St. Helena, which only broadcasts on shortwave one day a year, October 14 or 15, to celebrate St. Helena Day. Their transmission, on 11.0925 MHz, is a big event in the shortwave community. The announcers take phone calls from listeners, play music, and talk about their small island nation. Radio St. Helena receives over a thousand letters requesting station cards and souvenirs after every annual broadcast.

Q. How much power do shortwave stations run?

2-16 The Horizon Sales catalog carries Radio Caroline merchandise. Horizon Sales

A. From a few hundred watts to over a million watts. It all depends on the audience they want to reach and their budget. In parts of the world, where shortwave is more popular than it is in the United States, small, low power commercial stations that run from a few hundred to a few thousand watts are common. They broadcast primarily for people within their own country and, due to their low power, they are difficult for listeners overseas to hear.

International broadcast stations, who hope to reach listeners far beyond their own country's borders, run anywhere from several thousand watts to about a million watts. Of course, the higher a station's power is, the easier they are to hear!

The transmitter power used and your distance from a station are the prime considerations used for determining how good a "catch" it is. In radio language, "DX" means distance, and hearing a low-power domestic shortwave station from the other side of the world is a much better "DX catch" than hearing a high-powered international broadcast station that booms in night after night from the same region.

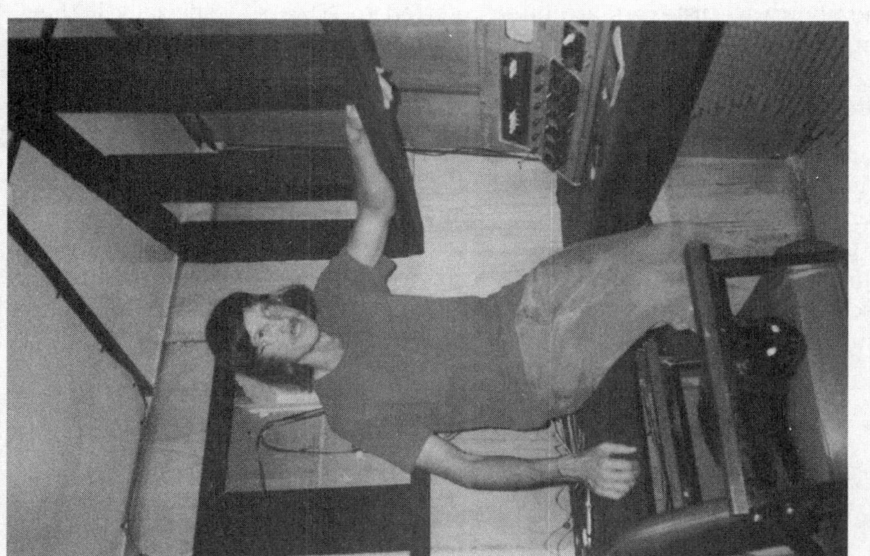

2-17 Allan Weiner of the *M/V Fury.*

Allan Weiner

2-18 Here's Radio New York International's Allan Weiner at the wheel of *The M/V Fury*.

In the shortwave hobby, you'll run into the term "DX" a lot. Many international shortwave broadcasters have "DX programs," where they read and discuss listening tips from hobbyists around the world. Shortwave clubs are often referred to as "DX clubs." And shortwave listeners are often referred to "DXers," especially if they spend most of their listening time trying to tune in low-power "domestic" stations!

Q. Why are there more programs in English at night than there are in the daytime?

A. Since most shortwave broadcasters have limited budgets, they broadcast their programs in English to North America at night, when more people are home and they are sure to have their largest audience. But, if you tune around, you are sure to find some programming in English, no matter when you have time to listen. The Voice of America, the British Broadcasting Corporation, Radio Canada International, Radio Moscow, HCJB Ecuador, and several other powerful stations can be heard broadcasting in English most any time of the day or night.

3
Communicating with shortwave stations

Q. How can I communicate with shortwave stations?

A. Most people communicate with shortwave stations by letter. It is the least expensive way to contact stations, and letters sent by air mail normally arrive at almost any location in the civilized world in about a week. In this age of global telecommunications you also have the option of communicating with most international broadcasters by telephone or fax.

Q. Which method is best?

A. It all depends on (1) your budget and (2) your reason for wanting to contact the station. In most instances, writing a letter and sending it air mail is the best way to go. But if money is no object, or you urgently want to get your message across, you can call long-distance or send a fax. Most international broadcast stations occasionally give their phone or fax number over the air. You can also check to see if their number is listed in the *World Radio-TV Handbook* or *Passport to World Band Radio*.

Some shortwave enthusiasts occasionally call or fax a hot listening tip into their favorite DX programs. That way, they can make sure that timely information arrives while it is still useful.

Several shortwave stations host call-in shows on special occasions, such as an anniversary or holiday. Radio For Peace International, for example, has a toll-free anniversary call-in program every September 16. This gives listeners the opportunity to say "hello" to their favorite hosts and have their voices heard on the international airwaves.

Q. What do they want to know?

A. Shortwave broadcasters want to know three things: who you are, what you think of their programming, and how well you can hear them.

Of course, telling a shortwave station who you are involves more than giving them your name and address! If they are to plan and air programs that you will enjoy, they need to know about your interests, age, occupation, or anything else you'd like to share.

Most people include something in their letter about their radio, their antenna system (if they have one), and how long they've been involved in the hobby.

Don't be ashamed to let them know if you are new to shortwave listening. Shortwave stations are always glad to welcome new listeners into the hobby, and often give them preference in getting their letters read on the air. They'll be glad to know you heard their broadcast and enjoyed it enough to write in and tell them about it!

Q. What is the best way to start your letter?

A. Just say "Hello, I'm ______, and I enjoy listening to your program." Then tell them some things about yourself, like your age, what you do for a living, the kind of community you live in, your other interests, how you got into shortwave listening, and what you like about their station. If you want, you can ask questions about the station or the country, or even suggest some programming ideas.

Q. Do shortwave stations really pay attention to listener's mail?

A. You bet they do! Broadcasters rely heavily on listener's comments to find out who is listening and what they want to hear. Do you want more news, editorials, environmental reports, business news, music, travel features—or would you rather hear something else entirely? It only takes a few minutes of your time to write and let them know. If you make your letter interesting enough, there's a good chance you can hear it read on an upcoming "Mailbag" show!

Q. Will shortwave stations send me anything in return?

A. Yes, especially if you request it. Nearly all shortwave stations have souvenir items to send out to their listeners. Station pennants (Fig. 3-1), photos, stickers, station post cards, and foreign stamps (Fig. 3-2) are among the most common items. Stations also send out program schedules so you'll know when to tune them in. Figure 3-3 shows schedules from Radio Sweden, Radio For Peace International, Costa Rica, and Pan-American Radio.

If you send them a report on how well their signal is reaching your area, they'll be happy to send you a QSL card.

Q. What's a QSL card?

A. In ham radio, QSL means "I verify reception." In other words, "Yes, you heard me!" So a QSL is a card or letter from a shortwave station that confirms you heard them. Nearly all shortwave listeners enjoy collecting QSL cards. One side of the card usually verifies your reception, while the other is decorated with photos of popular program hosts, the station logo (Fig. 3-4), scenic views, maps of the country (Fig. 3-5), or even antenna towers (Fig. 3-6).

Q. How do I get one?

3-1 Pennants from shortwave stations are fun to collect.

A. If you'd like to have a QSL card, you'll need to write what is known in the world of radio as a reception report.

A reception report should tell the station:

- What radio and antenna you use
- The time and date you heard the program
- The frequency (many shortwave stations broadcast on several frequencies at once, and want to know which one is being heard best in your part of the world)
- How well the signal could be heard in your area
- Some details of the program—to prove that you actually heard their station.

Shortwave reception conditions are usually reported by using the SINPO system. It is accepted and understood by shortwave listeners and station engineers worldwide. SINPO means:

S—Signal strength
I—Interference (QRM)
N—Noise (QRN)
P—Propagation (fading)
O—Overall quality

Q. How do I use the SINPO rating system?

A. Each category in the SINPO system is rated from 1 to 5, with 1 being the poorest reception conditions and 5 being the best.

A "1" signal strength rating means that you can barely hear the station, while a

3-2 Stamp collecting can add another dimension to your shortwave listening hobby.

"5" signal rating means you can hear it as well as if it were local. However, most of the shortwave transmissions you'll hear fall somewhere between the extremes, so you'll need to rate their signal as 2, 3, or 4.

The interference rating is almost as important as the signal strength rating to station engineers. A low rating, such as 1 or 2, means that another broadcaster on a nearby frequency is coming in on top of their transmission, and making it VERY difficult for you to hear them! Ratings of 3 or 4 mean that there is some interference but you can still hear most of the program, and a rating of 5 means that there is no interference at all. If you can identify the station that is causing interference, you should always include that information in your report. This information helps stations place themselves on the best possible frequencies for your listening pleasure.

3-3 Shortwave stations publish schedules of their transmissions.

3-4 A card from Radio HCJB, Ecuador.

Electrical noise can come from many sources: thunderstorms, flourescent lights, computers, power tools, and so on. Electrical noise is usually more severe on lower frequencies (closer to the AM band) than the higher ones (15 MHz and above). Extremely severe noise should be rated as 1, a complete absence of noise should be rated as 5, and anything between should be rated 2, 3, or 4.

Shortwave stations have to rely on the ionosphere, which can be quite unstable at times, to deliver their programs to listeners overseas. If there is a lot of fading during the show, rate the propagation as 1 or 2. If there is moderate fading, rate it 3 or 4. If the signal experiences no fading during the time of your report, rate it 5.

The overall rating in the SINPO system is just what it sounds like, an overall rating of signal quality. To rate the overall reception conditions, take all of the previous factors into consideration. Low ratings mean poor overall reception, and a high rating means good overall reception.

While station personnel are always happy to know that their overseas listeners can hear them loud and clear, you should never over-rate the SINPO on a reception report. Station engineers need to know how their signal is doing so they can find ways to improve reception in your area. And you're just as likely to receive a QSL card, win a contest, or get your letter read on the air whether you give their signal a high or low SINPO rating.

Figure 3-7 shows a reception report for a broadcast from Radio for Peace International, in Costa Rica.

3-5 This map card came from Radio Belize.

Q. How do I figure the correct time?

A. For many years, shortwave listeners and ham radio operators have used a world-wide standard time for their reception reports: UTC, or Universal Coordinated Time, is often announced during shortwave programs, especially at the top of the hour.

Standard time stations around the world broadcast the UTC time 24 hours a day. WWV, in Ft. Collins, Colorado, broadcasts the UTC time on 5, 10 and 15 MHz. WWV

3-6 An antenna card from Spain.

can easily be heard all over North America. A number of foreign countries also operate standard time stations.

In addition to telling you the right time, most of these stations will be happy to send you a QSL card for a correct reception report. You can find the address of any standard time station in the *World Radio-TV Handbook*. Figure 3-8 is a QSL card from the Australian time station VNG.

24-hour clocks, available through many shortwave mail-order companies, make it easy to know UTC without having to re-tune your radio. Figure 3-9 shows a standard 24-hour wall clock. Figure 3-10 is a 24-hour world map clock from MFJ Enterprises. Time wheels, which show you the correct time in any part of the world (Fig. 3-11), are also available.

Shortwave magazines, such as *Monitoring Times and Popular Communication*, list the time of English-speaking transmissions in both UTC and our local times. This is helpful to both new and experienced hobbyists. Table 3-1 is a UTC time conversion chart for North American listeners.

Q. Where can I get reception report forms?

Sep. 15, 1993

Dear Radio For Peace International,

I heard your 6th Anniversary call-in show tonight, and would very much like to have a RFPI QSL card.

Here's my reception report:

Time - 01:00 - 02:00 UTC Freq - 7.375 MHZ
Date - Sep. 15 1993 (Sep. 16 UTC date)
SINPO - 34433
Program Details:
Your announcers took phone calls from well-wishers all over the world. Fans called from Germany, California, West Virginia, Toronto and many other places.

Good luck with your station!

3-7 A reception report for Radio for Peace International, Costa Rica.

A. You can make your own reception report forms, or you can purchase them from some radio clubs or shortwave mail-order catalogs. If you are sending a report for only one or two broadcasts, you really don't need forms. But if you are sending several reports for programs broadcast from the same station, forms help you to better organize the information, making it easier for the station engineer to read and use.

Q. Do I need to send return postage?

3-8 A QSL card from VNG, the Australian standard time station.

3-9 A 24-hour wall clock.
Grove Enterprises, Inc.

A. Most international broadcast stations have a budget for sending out QSL cards, pennants, and other items. But when you write to regional or domestic (overseas) shortwave stations, it is always a good idea to send return postage in the form of two or three international reply coupons, also known as IRCs. International Reply Coupons are available at your local post office, and can be exchanged for postage stamps almost anywhere in the world.

The World Radio-TV Handbook gives the QSL policy of most of the stations they list. If you are in doubt as to whether a station you'd just love to have a QSL

3-10 The MFJ DXer's World Map Clock shows you the time anywhere in the world.

MFJ Enterprises, Inc.

from will pay the return postage, it's a good idea to put a few IRCs in with your reception report.

When you write to AM stations within your own country, you can improve your chances of getting a reply by enclosing a self-addressed stamped envelope, or SASE.

When you write to pirate stations, always enclose three mint first class stamps for return postage. One stamp will take it from the maildrop to the pirate, the second stamp will take it from the pirate back to the maildrop, and the last stamp will bring it from the maildrop to your mailbox! (Fig. 3-12).

Figure 3-13 is a QSL from the pirate station Radio U.S.A.

Q. How long does it take to hear from a shortwave station?

A. It depends. Sometimes you'll hear from a station within a few weeks, but often it can take months. An airmail letter should reach nearly any destination in the world in about a week, but with budget cutbacks most shortwave stations don't have the staff they would need to get a reply back to you as quickly as you would like. On average, most shortwave stations will answer your request for a QSL card within a few months of receiving it. Just keep writing to all the stations you enjoy, and you will eventually be rewarded with a mailbox full (Fig. 3-14) of beautiful QSL cards!

Q. How many letters do shortwave stations receive?

A. Many shortwave stations—especially the Voice of America, The British Broadcasting Corporation, Deutsche Welle (The Voice of Germany), Radio Moscow, and Radio Canada International (Fig. 3-15), receive hundreds of letters a week.

Q. Do AM stations send out souvenir items?

A. The bigger the station is, the more likely they are to have a budget for sending out souvenir items. But it never hurts to ask; some small stations do send nice souvenirs to out-of-town listeners who take the time to let the stations know they heard them. QSL cards, bumper stickers, and photos of program hosts are some of the most requested items. Figure 3-16 is a card from Radio WABC 77 AM, New York, NY.

Grove Enterprises, Inc.

3-11 A World View Time Wheel.

Table 3-1.
UTC time conversion for North American listeners

UTC	EST	CST	MST	PST
0000	7:00 PM	6:00 PM	5:00 PM	4:00 PM
0100	8:00 PM	7:00 PM	6:00 PM	5:00 PM
0200	9:00 PM	8:00 PM	7:00 PM	6:00 PM
0300	10:00 PM	9:00 PM	8:00 PM	7:00 PM
0400	11:00 PM	10:00 PM	9:00 PM	8:00 PM
0500	Midnight	11:00 PM	10:00 PM	9:00 PM
0600	1:00 AM	Midnight	11:00 PM	10:00 PM
0700	2:00 AM	1:00 AM	Midnight	11:00 PM
0800	3:00 AM	2:00 AM	1:00 AM	Midnight
0900	4:00 AM	3:00 AM	2:00 AM	1:00 AM
1000	5:00 AM	4:00 AM	3:00 AM	2:00 AM
1100	6:00 AM	5:00 AM	4:00 AM	3:00 AM
1200	7:00 AM	6:00 AM	5:00 AM	4:00 AM
1300	8:00 AM	7:00 AM	6:00 AM	5:00 AM
1400	9:00 AM	8:00 AM	7:00 AM	6:00 AM
1500	10:00 AM	9:00 AM	8:00 AM	7:00 AM
1600	11:00 AM	10:00 AM	9:00 AM	8:00 AM
1700	Noon	11:00 AM	10:00 AM	9:00 AM
1800	1:00 PM	Noon	11:00 AM	10:00 AM
1900	2:00 PM	1:00 PM	Noon	11:00 AM
2000	3:00 PM	2:00 PM	1:00 PM	Noon
2100	4:00 PM	3:00 PM	2:00 PM	1:00 PM
2200	5:00 PM	4:00 PM	3:00 PM	2:00 PM
2300	6:00 PM	5:00 PM	4:00 PM	3:00 PM

For Daylight Savings Time, look ahead one hour on UTC column.

Q. Does it take very long to hear from AM stations in the United States or Canada?

A. Sometimes yes, sometimes no. Since AM stations are operated mostly to serve a local audience, they don't normally have a staff to check reception reports, answer listener's questions, and send out souvenirs. That job is usually handled by a secretary or one of the station engineers, in his or her spare time. While some stations are fairly prompt about answering listener's mail, others put it at the bottom of their "to do" list. So again, you might have to wait awhile for that sought-after QSL card. Figure 3-17 shows QSL cards from WBB Chicago, IL, and WHAS Louisville, KY. Figure 3-18 is a verification letter from Radio KBIS, in Little Rock, Arkansas.

Q. Are there ways other than QSL cards to prove I heard a station?

A. Yes. Some long-distance listeners enjoy recording on cassette tape the stations they hear. If you can't find time to write letters or the postage bill is getting too high, cassette recordings can be an alternative way to document your listening accomplishments. Even if you do write in for QSL cards, recordings of your favorite programs and rare DX catches are fun to hear over and over again.

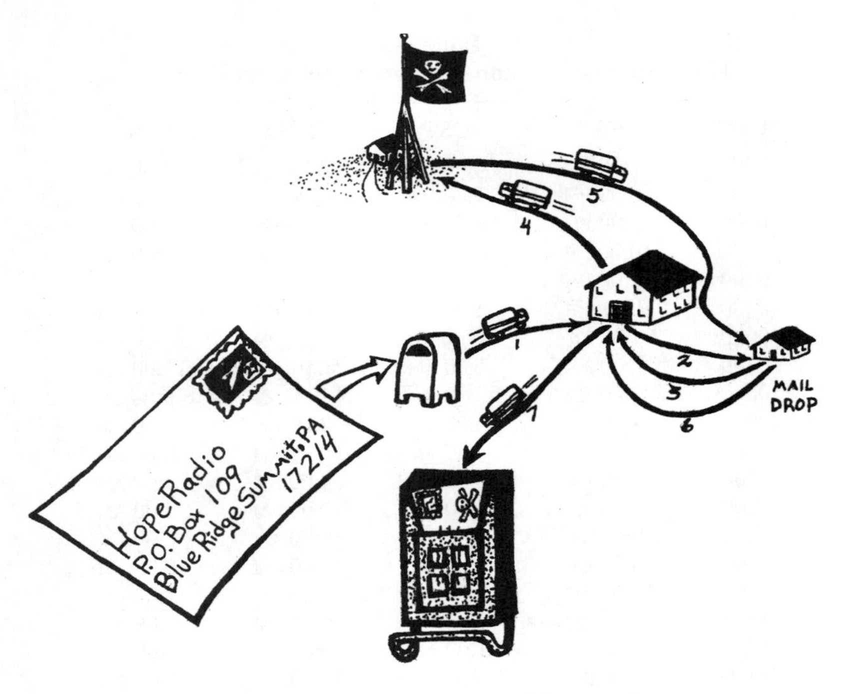

3-12 This is how a pirate maildrop works.

3-13 Here's a verification from pirate station Radio USA.

3-14 A mailbox full of colorful QSL cards is well worth the wait!

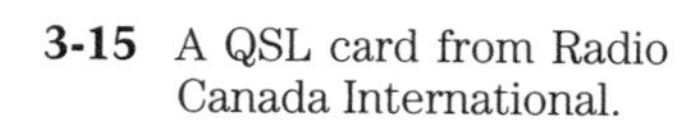

3-15 A QSL card from Radio Canada International.

3-16 Here's a card from WABC's "Angels in the Morning" show.

3-17 QSL cards from AM stations WBBM, Chicago, and WHAS, Louisville.

2400 Cottondale Lane • Little Rock, AR 72202 • 501-664-9410 • FAX: 501-664-5871

03/18/93

Anita Louise McCormick

This letter will serve to confirm your reception of KBIS-AM at 1010 kHz on 03/11/92 at 7:00pm CST. The programming details you described do reflect an accurate representation of what was broadcast at that time.

The KBIS facility is licensed for non-directional operation at 10,000 watts from local sunrise to local sunset. During the nighttime we transmit a directional signal at 5,000 watts utilizing a four tower array. The nighttime pattern is an asymetrical figure-8 oriented roughly along a line ENE to WSW.

Please accept my apology for the long delay in answering your letter. Unfortunately, circumstances in recent past have not allowed time for me to answer the many DX reports that I have received such as yours. I do hope this letter will serve your purposes.

Thanks for your report and good luck in your hobby.

Sincerely,

Norm Laramee
Chief Engineer KBIS-AM

3-18 A verification letter from Radio KBIS, Little Rock, Arkansas.

4
Hearing hams and utility stations

Q. How can I hear ham radio operators?

A. You can hear ham radio operators from around the world on any shortwave radio that has a SSB (single sideband) control (without a SSB control most ham transmissions sound garbled). And you can hear ham radio operators living within several miles of you on almost any scanner.

Ham radio operators come from all walks of life. When you tune across the ham bands, you never know who you might hear. It could be a teacher, electronic engineer, mailman, chemist, writer, minister, farmer, high school student, airline pilot, even an astronaut (Fig. 4-1)!

The subjects you'll hear discussed are almost limitless. While some technically inclined operators like nothing better than to endlessly discuss their antennas and transmitting equipment, many hams enjoy engaging other operators in on-the-air discussions on topics ranging from ecological issues, to plans for remodeling the living room, or even politics.

Q. Which frequencies do ham radio operators use?

A. Ham radio operators use several groups of frequencies in the radio frequency spectrum. They range from 1,800 kHz (just above the AM band) to 10GHz (in the microwave band). Figure 4-2 is a chart prepared by the American Radio Relay League, showing the location of the ham radio bands, and telling which class of license and which type of transmission can be used in each section of the band. Some sections are reserved for phone (voice), some are reserved for Morse Code (abbreviated CW), and other sections are for both Morse Code and radioteletype (RTTY) operation.

The American Radio Relay League has a ham station of its own, W1AW, located at their headquarters in Newington, CT. W1AW broadcasts bulletins of news and developments in the amateur radio service in both voice and radioteletype. They also

4-1 Astronaut Katheryn D. Sullivan at the Johnson Space Center Amateur Radio Club station.

broadcast code practice sessions at speeds ranging from 5 to 35 words per minute, using material taken from their monthly magazine, *QST*. Figure 4-3 is a schedule of W1AW's daily transmissions.

Q. Why do ham radio operators use Q-signals, abbreviations, and phonetics?

A. Hams use Q-signals (Table 4-1) and abbreviations (Table 4-2) because they save time; radio operators around the world understand their meaning. In the early years of radio, when nearly all hams used Morse Code, Q-signals and abbreviations quickly gained popularity. Since operators using these shortcuts didn't have to send so many dits and dahs, they could get the message out more quickly!

Phonetics (Table 4-3) came into use when ham radio operators started using voice communications. By spelling out their call letters, names, or other important information in phonetics, hams can make it easier for other operators to hear him or her when band conditions are bad. Phonetics can make the difference between getting a QSL card from a ham in an exotic new location, or having the guy on the other end say "Sorry, I can't quite make out your call letters!"

Q. Do hams have QSL cards?

A. Yes. Nearly all hams have QSL cards. Ham QSLs show the operator's call letters, name, and address. Figure 4-4 is a card from K3CR, at the Penn State Amateur Ra-

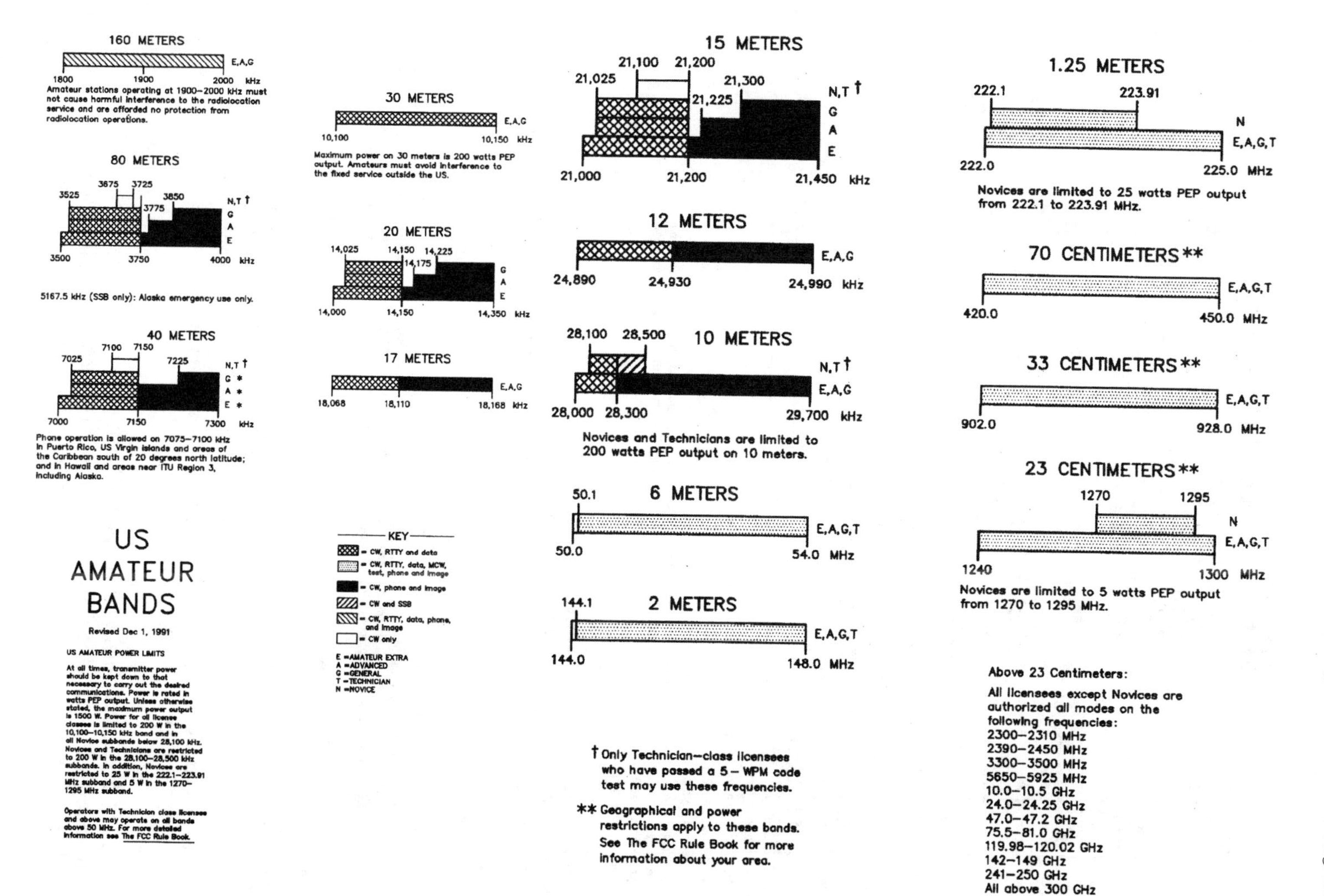

4-2 This chart shows where amateurs in the USA can operate. American Radio Relay League

W1AW schedule

Pacific	Mtn	Cent	East	Sun	Mon	Tue	Wed	Thu	Fri	Sat
6 am	7 am	8 am	9 am			Fast Code	Slow Code	Fast Code	Slow Code	
7 am	8 am	9 am	10 am			Code Bulletin				
8 am	9 am	10 am	11 am			Teleprinter Bulletin				
9 am	10 am	11 am	noon							
10 am	11 am	noon	1 pm							
11 am	noon	1 pm	2 pm		Visiting Operator Time					
noon	1 pm	2 pm	3 pm							
1 pm	2 pm	3 pm	4 pm	Slow Code	Fast Code	Slow Code	Fast Code	Slow Code	Fast Code	Slow Code
2 pm	3 pm	4 pm	5 pm	Code Bulletin						
3 pm	4 pm	5 pm	6 pm	Teleprinter Bulletin						
4 pm	5 pm	6 pm	7 pm	Fast Code	Slow Code	Fast Code	Slow Code	Fast Code	Slow Code	Fast Code
5 pm	6 pm	7 pm	8 pm	Code Bulletin						
6 pm	7 pm	8 pm	9 pm	Teleprinter Bulletin						
6[45] pm	7[45] pm	8[45] pm	9[45] pm	Voice Bulletin						
7 pm	8 pm	9 pm	10 pm	Slow Code	Fast Code	Slow Code	Fast Code	Slow Code	Fast Code	Slow Code
8 pm	9 pm	10 pm	11 pm	Code Bulletin						
9 pm	10 pm	11 pm	Mdnte	Teleprinter Bulletin						
9[45] pm	10[45] pm	11[45] pm	12[45] am	Voice Bulletin						

❑ ***Morse Code transmissions:***

Frequencies are 1.818, 3.5815, 7.0475, 14.0475, 18.0975, 21.0675, 28.0675 and 147.555 MHz.

Slow Code = practice sent at 5, 7-1/2, 10, 13 and 15 WPM.

Fast Code = practice sent at 35, 30, 25, 20, 15, 13 and 10 WPM.

Code practice text is from the pages of *QST*. The source is given at the beginning of each practice session and alternate speeds within each session. For example, "Text is from July 1992 *QST*, pages 9 and 81" indicates that the plain text is from the article on page 9 and mixed number/letter groups are from page 81.

Code bulletins are sent at 18 WPM.

❑ ***Teleprinter transmissions:***

Frequencies are 3.625, 7.095, 14.095, 18.1025, 21.095, 28.095 and 147.555 MHz.

Bulletins are sent at 45.45-baud Baudot and 100 baud AMTOR, FEC Mode B. 110-baud ASCII will be sent only as time allows.

On Tuesdays and Saturdays at 6:30 PM Eastern time, Keplerian Elements for many amateur satellites are sent on the regular teleprinter frequencies.

❑ ***Voice transmissions:***

Frequencies are 3.99, 7.29, 14.29, 18.16, 21.39, 28.59 and 147.555 MHz.

❑ ***Miscellanea:***

On Fridays, UTC, a DX bulletin replaces the regular bulletins.

W1AW is open to visitors during normal operating hours: from 1:00 PM until 1 AM on Mondays, 9 AM until 1 AM Tuesday through Friday, and from 3:30 PM to 1 AM on Saturdays and Sundays. FCC licensed amateurs may operate the station from 1-4 PM Monday through Friday. Be sure to bring your current FCC amateur license or a photocopy. Special arrangements must be made for weekend operation; please call or write at least a week in advance.

In a communications emergency, monitor W1AW for special bulletins as follows: voice on the hour, teleprinter at 15 minutes past the hour, and CW on the half hour.

Headquarters and W1AW are closed on New Year's Day, President's Day, Good Friday, Memorial Day, Independence Day, Labor Day, Thanksgiving and the following Friday, and Christmas Day. On the first Thursday of September, Headquarters and W1AW will be closed during the afternoon.

4-3 The American Radio Relay League station, W1AW, provides code practice sessions and bulletins of ham radio news. American Radio Relay League

dio Club. Ham radio operators usually exchange cards with other hams they contact on the air. But many hams will send QSLs to shortwave listeners, especially if they are thoughtful enough to enclose a stamped, self-addressed envelope.

Table 4-1. International Q signals

QRL	Is this frequency busy?
QRM	Interference from other stations
QRN	Atmospheric noise (static)
QRO	Increase transmitter power
QRP	Lower transmitter power
QRS	Send (code) slower
QRT	Stop transmitting
QRZ	What are your call letters?
QSA	What is my signal strength?
QSB	Your signal is fading
QSL	Did you copy? (also means please send a QSL card)
QSO	Contact (2-way transmission)
QSY	Change frequency to ______
QTH	Location (usually city and state)

Table 4-2. Abbreviations for Morse Code operators

ABT	About
ADR	Address
AGN	Again
ASCII	American National Standard Code for Information Interchange
BCI	Broadcast (shortwave) interference
BCNU	Be seeing you (good by)
BK	Back—or—break-in
C	Yes
CFM	Confirm
CL	Closing, signing off the air
CQ	Calling any station
CUL	See you later
CW	Continuous Wave—Morse Code
DR	Dear (used as a greeting by many overseas hams)
DX	Distance
ES	And
FB	Fine business (great)
FREQ	Frequency
GA	Good afternoon—or—go ahead
GE	Good evening
GM	Good morning
GN	Good night
HI	Laughter in CW
HR	Here
HV	Have
HW	How
ITU	International Telecommunications Union
LID	A bad operator

Table 4-2. Continued.

N	No
NCS	Net control station
NR	Number
NW	Now
OM	Old man—a way of referring to another operator until you know his name.
PSE	Please
R	Received
RCVR	Receiver
RFI	Radio frequency interference
RPT	Repeat
RTTY	Radio teletype
SASE	Self-addressed stamped envelope
SIG	Signal
SKED	Schedule
SRI	Sorry
SSB	Single side band
SWL	Shortwave listener
TNX	Thanks
TVI	Television interference
VFO	Variable frequency oscillator
VY	Very
WA	Word after
WB	Word before
WPM	Words per minute
WX	Weather
XMTR	Transmitter
XTAL	Crystal
XYL/YF	Wife
YL	Single lady
73	Best wishes
88	Love and kisses

**Table 4-3.
ITU (International
Telecommunications Union) phonetics**

A Alfa	**I** India	**R** Romeo
B Bravo	**J** Juliett	**S** Sierra
C Charlie	**K** Kilo	**T** Tango
D Delta	**L** Lima	**U** Uniform
E Echo	**M** Mike	**V** Victor
F Foxtrot	**N** November	**W** Whiskey
G Golf	**O** Oscar	**X** X-ray
H Hotel	**P** Papa	**Y** Yankee
	Q Quebec	**Z** Zulu

4-4 A QSL card from the Penn State Amateur Radio Club.

Q. How do I write a reception report to a ham station?

A. Hams use a different reporting system than shortwave stations. Instead of using the SINPO system, they use the RST(Readability, Signal, Tone) system. When you are reporting on voice transmissions, you only need to rate the readability and signal. Table 4-4 shows how the RST system works. Notice that Readability ratings run from 1 to 5, while Signal and Tone ratings run from 1 to 9.

Table 4-4. The RST signal reporting system

Readability

1 Unreadable
2 Barely readable
3 Readable with considerable difficulty
4 Readable with practically no difficulty
5 Perfectly readable

Signal Strength

1 Faint signal - barely perceptible
2 Very weak signal
3 Weak signal
4 Fair signal
5 Fairly good signal
6 Good signal
7 Moderately strong signal
8 Strong signal
9 Extremely strong signal

Tone (for code transmissions only)

1 Extremely rough, harsh, and broad tone
2 Very rough, harsh tone

Table 4-4. Continued.

3 Rough, rippling tone
4 Moderately rough, rippling tone
5 Moderate, with some rippling sound
6 Moderate, hardly any rippling in tone
7 Near pure tone, only traces of rippling
8 Near perfect tone
9 Perfect tone

When you write reception reports to ham radio stations, be sure to include the date, time (in UTC), and the frequency the operators were using. Also, it's a good idea to include a few details of the conversation you monitored, just to prove that you actually heard the operator you're requesting a QSL from.

You can find ham radio operators addresses in the *Radio Amateur Callbook*. Both domestic and foreign editions are published every year. Many libraries have the *Radio Amateur Callbook*—or you can order them through ham/shortwave mail-order companies.

Q. What is a utility station?

A. The easiest way to define utility stations is to say that they are NOT on the air for the purpose of informing or entertaining the general public. They are there as a means of communication between individuals, services, companies, and countries.

Utility stations and their services cover an extremely wide range. They are weather stations that provide forecasts for ships and planes, military transmissions of all kinds, embassy communications, ship-to-ship and ship-to-shore communication (Fig. 4-5), press service transmissions, phone calls from passengers on ships at sea, police dispatches, airline pilots communicating with the "ground control" personnel at airports, and too many other things to mention.

Q. Where can I hear utility transmissions?

A. Utility stations broadcast on frequencies throughout the shortwave and scanner bands, spaced between the "broadcast stations" that serve the general public. *Monitoring Times*, *Popular Communications,* and a number of DX clubs carry monthly columns on utility transmissions.

Many frequency guide books are available to help you find the utility stations you're most interested in hearing. The *International Callsign Handbook*, by Gale Van Horn (Fig. 4-6), is a good reference book for anyone interested in tuning in utility stations. This 250-page directory lists callsigns from around the world and identifies the stations using them on both shortwave and scanner frequencies.

If you're interested in aeronautical utility communications, you'll want to pick up the *Official Aeronautical Directory,* (Fig. 4-7), available at Radio Shack stores nationwide. It contains information on aviation communications on all bands. Radio Shack also sells the *Police Call Radio Guide,* (Fig. 4-8), filled with information on

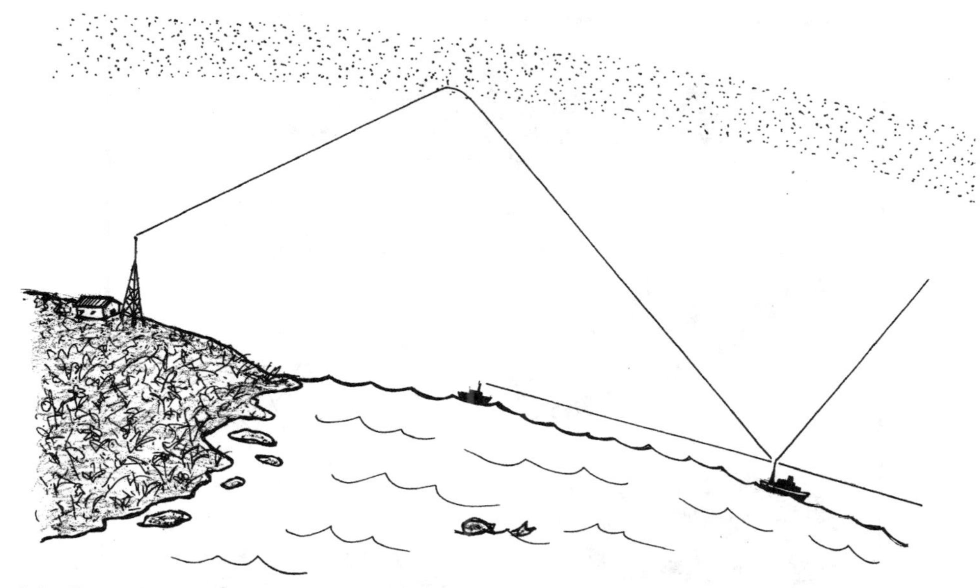

4-5 Ships communicate with other ships and land-based stations on the shortwave radio bands.

police, fire department, and other emergency communications. Both books are also available through shortwave/scanner mail-order dealers.

In order to save space, most utility columns and guide books use abbreviations. Table 4-5 lists some of the most commonly used utility abbreviations.

Q. What do police 10 codes mean?

A. Unfortunately, not all police departments assign the same meaning to the 10 codes. Table 4-6 is the official Association of Police Communications Officers 10 code list. If the codes on this list don't seem to be what law enforcement officers in your area are using, ask local scanner hobbyists if they have a copy of the correct list for your area's department. Electronics stores and department stores that sell scanners might also be able to help you.

Q. What equipment do I need to hear utility stations?

A. As with ham radio stations, any shortwave radio with a single sideband control or any scanner (sometimes known as a "police radio") will bring you plenty of utility stations. If you want to be able to identify and read the Morse Code and radioteletype transmissions you hear, you should look into adding a multi-mode digital decoder to your listening post.

Q. What can I do with a digital decoder?

A. Digital decoders take the dits, dahs, chirps, and peeps you hear on the shortwave band, and translate them into the letters and numbers they represent. Many

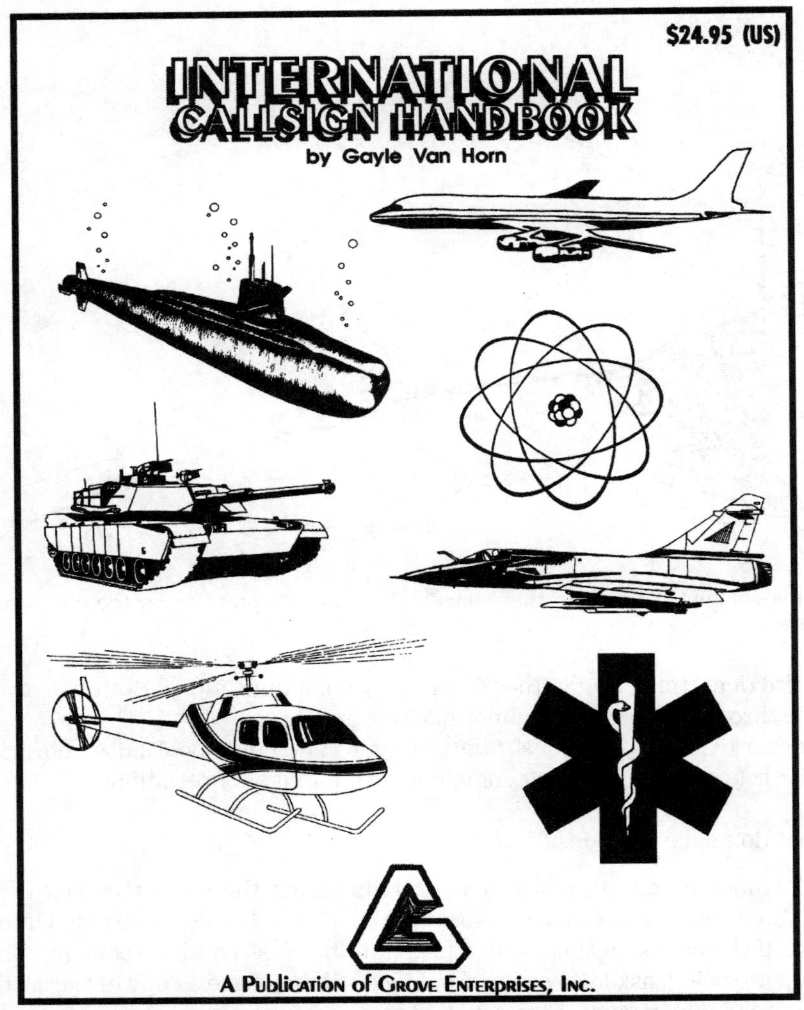

4-6 *The International Callsign Handbook* is an excellent reference book for utility listeners. Grove Enterprises, Inc.

ham and utility stations operate in digital modes, and owning a decoder will open up a whole new facet of the hobby to you. Several brands of digital decoders are available through the major shortwave mail-order companies, and each has its own translating capability.

One of the lowest-priced decoders on the market is the Somerset Microdec MD 100 multi-mode decoder, priced around $200. It translates Morse Code, standard RTTY (radioteletype), and the ASCII computer code. You can use it as a stand-alone unit (with any shortwave radio that has SSB reception)—or if you purchase a Somerset VIP50 video/printer interface, you can display what you've decoded on a computer monitor or television screen. Somerset Electronics, Inc., also manufactures

4-7 *The Official Aeronautical Frequency Directory.*
Tandy Corporation/Radio Shack

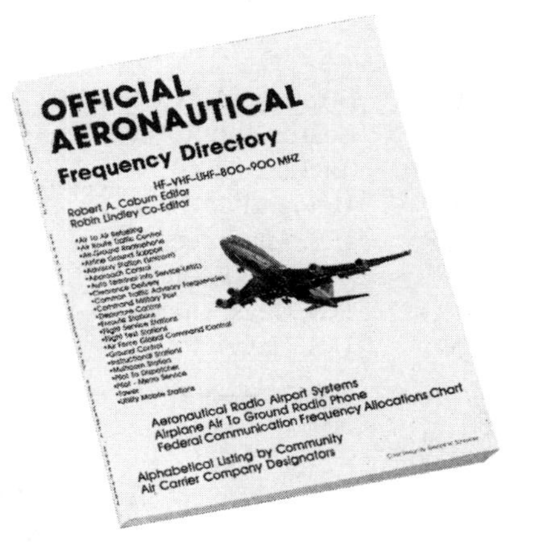

Tandy Corporation/Radio Shack

4-8 *The Police Call Radio Guide.*

Table 4-5. Common utility abbreviations

AFB	Air Force Base
ASCII	American Standard Code for Information Interchange (most commonly used code for exchanging computer data)
ATC	Air Traffic Control
CAP	Civil Air Patrol
CG	Coast Guard
CQ	General call any station may answer
CW	Continuous wave - Morse Code
DE	French word "from," used in CW and other digital communications
DOT	Department of Transportation
EAM	Emergency Action Message
FAA	Federal Aviation Administration
FEMA	Federal Emergency Management Agency
ID	Identification
LSB	Lower Side Band
MARS	Military Affiliate Radio System
M/V	Motor Vessel
Net	Network
RAF	Royal Air Force
RTTY	Radio Teletype

Table 4-5. Continued.

SAC	Strategic Air Command
SAR	Search and Rescue
TAC	Tactical
UNID	Unidentified
USAF	U.S. Air Force
USB	Upper Side Band
USCG	U.S. Coast Guard
USN	U.S. Navy

Table 4-6.
Association of Police Communications Officers 10-code

10-1	Signal weak
10-2	Signal good
10-3	Stop Transmitting
10-4	Affirmative (OK)
10-5	Relay (to)
10-6	Busy
10-7	Out of Service
10-8	In Service
10-9	Say Again
10-10	Negative
10-11	_______ on duty
10-12	Stand by (stop)
10-13	Existing conditions
10-14	Message/Information
10-15	Message delivered
10-16	Reply to message
10-17	Enroute
10-18	Urgent
10-19	(In) contact
10-20	Location
10-21	Call _________ by phone
10-22	Disregard
10-23	Arrived on scene
10-24	Assignment completed
10-25	Report to (meet)
10-26	Estimated arrival time
10-27	License/permit information
10-28	Ownership information
10-29	Records check
10-30	Danger/caution
10-31	Pick up
10-32	______ units needs (specify)
10-33	Help me quick
10-34	Time

4-9 The Microdec MD 300 and a Model VIP50 Video/Printer Interface.

more advanced decoders that are able to decode the newer radioteletype modes, such as AMTOR/SITCOR, that a growing number of utility stations now use. Figure 4-9 is a Microdec MD 300 with a VIP150 Video/Printer Interface.

Universal Radio has a line of moderately priced decoders able to translate many kinds of digitally transmitted data. The Universal M-400 Decoder (Fig. 4-10), priced at under $450, decodes Morse Code, RTTY, and many of the new radioteletype modes. You can read the decoded text on the M-400's 40-character readout display. The M-400 also decodes FAX transmissions of weather maps, which you can print out on a parallel computer printer.

4-10 The Universal M-400 Decoder.

The Universal M-1200 Decodercard (Fig. 4-11), available in the same price range as the M-400, fits into an expansion slot in your computer. It decodes even more modes of digital transmissions than the M-400, displaying text and pictures on your computer screen. With the M-1200, you also have the option of saving the data you decode on computer disks for future reference.

4-11 The Universal M-1200 Decodercard.

No matter what kind of multi-mode digital decoder you own, there will always be some digital transmissions that you cannot decode. These transmissions, often containing secret information, are sent in a code that (hopefully) only the intended recipient can decipher!

5
Radio waves—how they make the trip

Q. Just what is a *radio wave*?

A. A radio wave is an electromagnetic field. Radio waves are created at the transmitter when electrical charges are switched from positive to negative and back again thousands of times a second. These waves of electromagnetic energy, containing the music, comedies, dramas, talk shows, and news programs we all enjoy, are fed through a transmission line to the antenna, where they are pushed off into the air . . . and begin their journey.

Figure 5-1 is a diagram of a radio wave. The distance from one energy peak to the next is the "wavelength," and the number of waves that pass any given point per second is the frequency (the number on your radio dial). If you tune your shortwave receiver to 7.415, MHz for example, you will be listening to radio waves that pass your antenna at the rate of 7.415 million waves per second!

Frequency charts (Fig. 5-2) illustrate the radio services that the Federal Communications Commission has assigned to various parts of the radio frequency spectrum.

Q. What is the difference between AM, FM, and SSB?

A. AM, FM, and SSB are methods of encoding information onto a radio wave.

AM (Amplitude Modulation) is one of the oldest methods of broadcasting voice and music over the airwaves. It has been in use since the early part of the century, and is still used on the domestic AM "broadcast band" as well as for most shortwave broadcasting. In AM transmissions, the amplitude, or the strength, of the "carrier" radio wave is varied to transport the two sidebands, containing identical programming, from the transmitter to your receiver.

In FM (Frequency Modulation) broadcasting, the frequency is varied in order to get the message from one place to another. FM has the dual advantages of higher

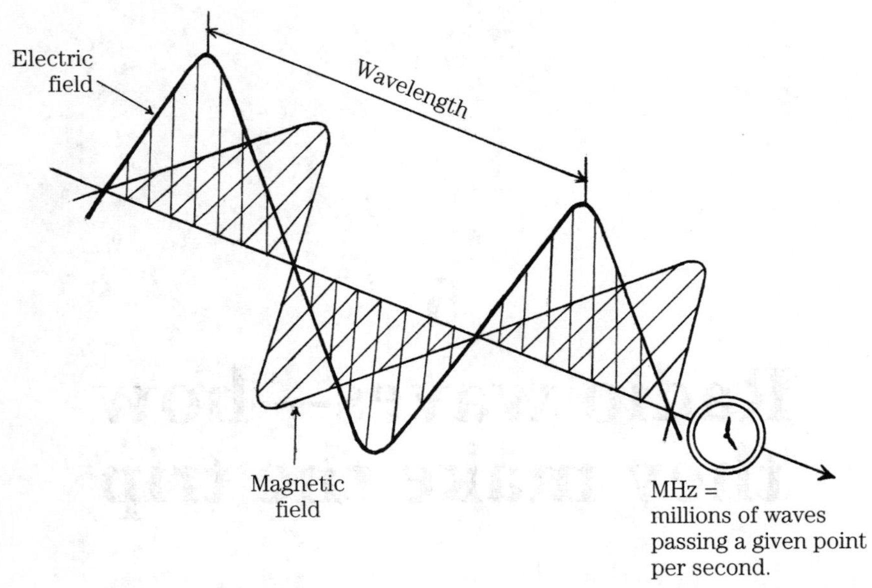

5-1 Radio waves are made up of an electric field and a magnetic field.

fidelity and a relative freedom from electrical noises and static, but it has the disadvantage of taking up more bandspace than AM transmissions. Frequency modulation is used in the 88–108 MHz band, and for some utility and scanner transmissions.

SSB (Single Side Band) is similar to AM, the amplitude is varied in order to encode information on radio waves. But in single sideband transmissions, the carrier is suppressed and one of the sidebands is eliminated before transmissions are put on the air. Ham radio operators, utility stations, and a few shortwave broadcasters use single sideband. If you want to hear SSB transmissions, you'll need a shortwave radio with a SSB control, otherwise the signal will sound garbled. When the SSB is switched on, it replaces the missing parts of the signal, and makes voice and music transmissions sound nearly normal. Although SSB has the advantage of occupying one half the bandwidth of an AM signal and requiring much less power to transmit messages or programming to the same location, it has the disadvantages of being harder to tune and requiring more expensive equipment to receive, and the sound quality of SSB is not as good as AM or FM.

Q. What causes radio signals to "skip"?

A. The ionosphere, (Fig. 5-3) which is made up of several electrically charged layers of gas high up in the atmosphere, makes it possible for you to hear great shortwave radio programs from overseas. When shortwave signals hit the ionosphere, they bend and are often reflected back down to Earth. The ionosphere is always changing, and each layer of the ionosphere has a different effect on the radio signals that reach it (Fig. 5-4).

Radio Frequency Spectrum

Between the very low frequencies, which can be heard as sound, and the very high frequencies which can be seen as light, are the very useful frequencies we use for radio communications. Although the exact limits are only roughly defined and subject to change as our technology advances, the most usable frequencies are clustered in the center of this radio spectrum. The bar graph below illustrates radio's position in the frequency spectrum.

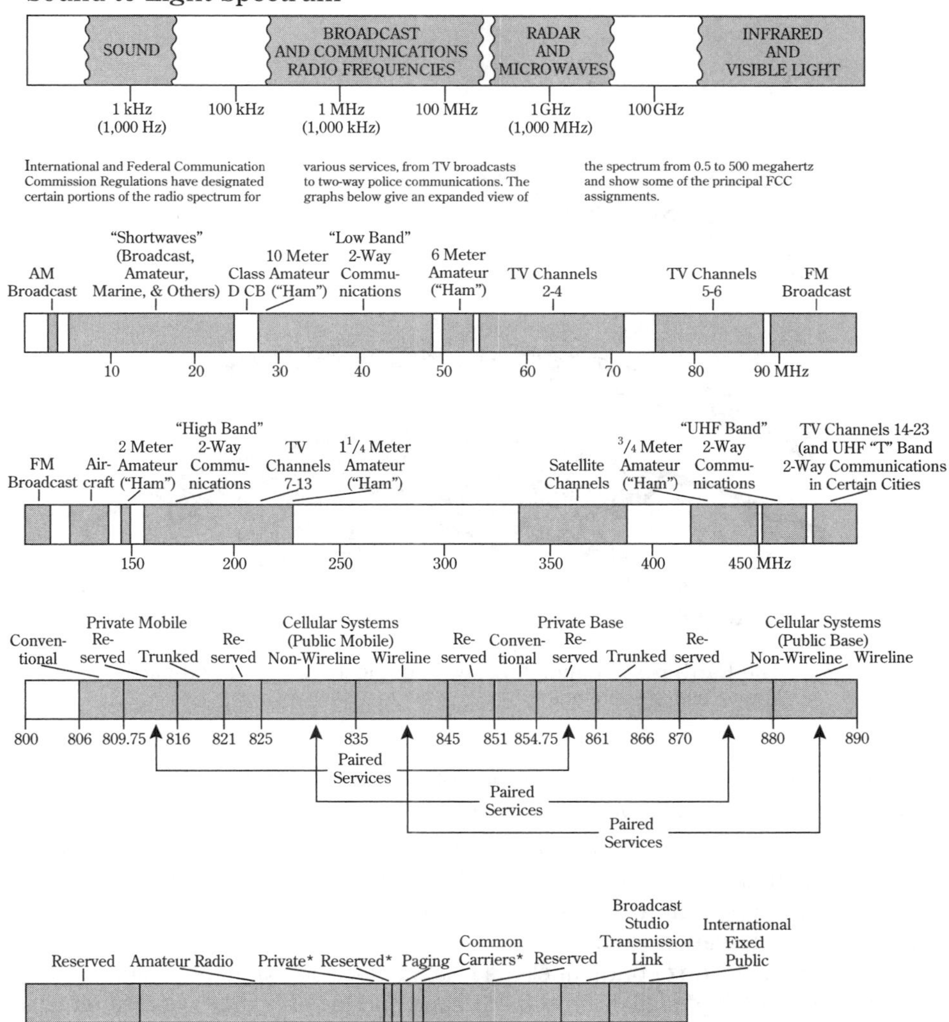

5-2 Radio frequency spectrum chart. Bearcat Radio Club, Kittering, OH

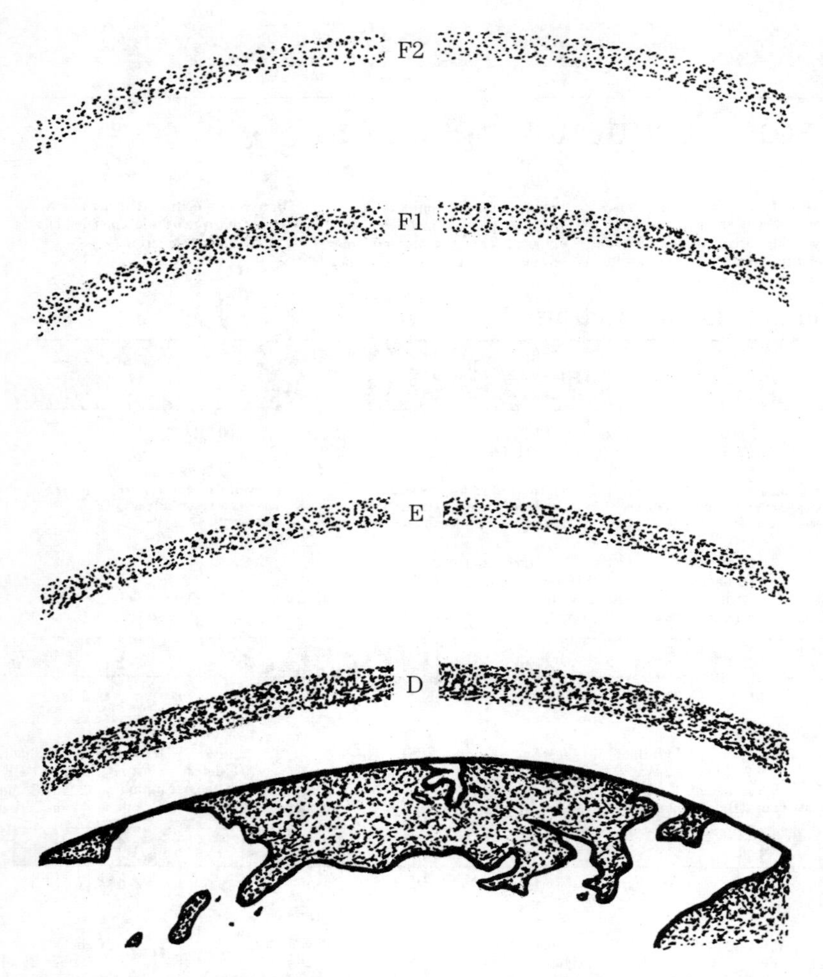

5-3 The Earth is surrounded by layers of electrically charged particles, known as the ionosphere.

Q. Why can I hear shortwave stations on some bands better in the daytime and have better reception on other bands at night?

A. During the day, radiation from the sun "charges up" the ionosphere above us. In the evening hours, when the side of the Earth we're on turns away from the sun, we move through areas of the ionosphere that are not directly exposed to the sun's endless stream of radiation (Fig. 5-5).

On the night side of our planet the D layer of the ionosphere, which absorbs and destroys radio "skip" signals on the AM band and lower shortwave frequencies during the day, does not exist. That is why you can hear AM stations hundreds of miles away at night, but only hear local stations during the day.

At night, the E layer of the ionosphere is thinner, and the F 1 and F 2 layers, merged, are hovering about 100 miles above us. Those changes make it possible to

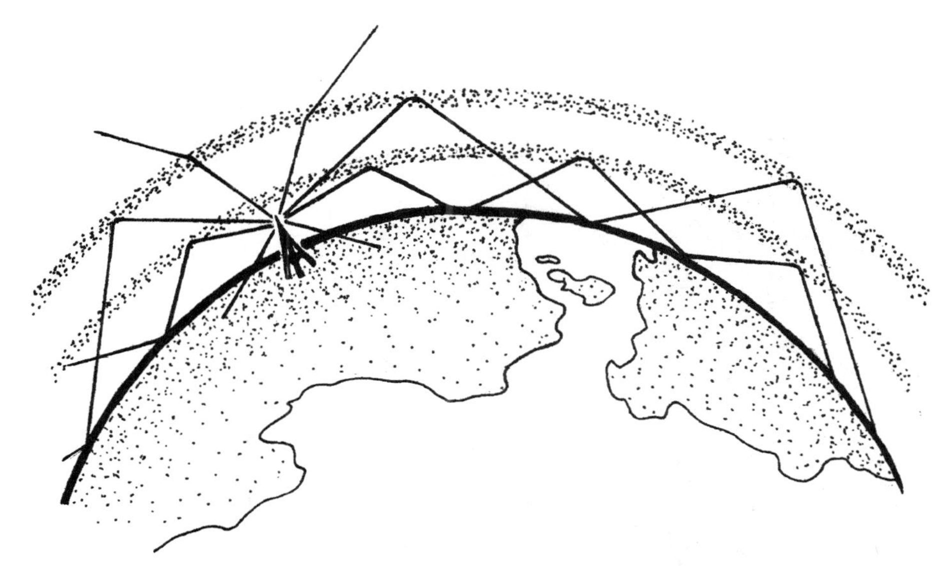

5-4 Each layer of the ionosphere has a different effect on radio skip conditions.

hear more distant low frequency stations at night, and more distant high frequency (15 MHz and above) stations during daylight hours.

Q. How do you use the propagation charts that appear in shortwave magazines?

A. Propagation charts give predictions of when you will have the best reception conditions from one part of the world to another. They chart the Maximum Useable Frequency (MUF), the Lowest Useable Frequency (LUF), and the Optimum Useable Frequency (OUF), which should give you the best reception of all! Propagation charts show the time of day in UTC on one side, and the frequency in MHz on the other.

Propagation charts published in North American magazines such as *Monitoring Times* usually give conditions between Eastern USA and other parts of the world, and Western USA and other parts of the world. Figure 5-6 is a propagation chart showing the best times and frequencies for a listener in Eastern USA to tune in Australia.

Propagation conditions are always changing. The time of day, season of the year, and the amount of radiation the ionosphere receives from the sun all influence which frequencies will give you the best reception.

Q. How does the sun affect radio wave propagation?

A. The sun goes through energy cycles, lasting roughly 11 years, with a peak and a drop in the amount of radiation it sends out. High radiation periods are characterized by a high number of sunspots (Fig. 5-7). At the high end of the solar cycle, there is very good reception of stations broadcasting on 15 MHz and above. During the low end of the cycle, when there is less solar radiation charging the ionosphere, distant stations on the AM band and lower shortwave frequencies come in much better.

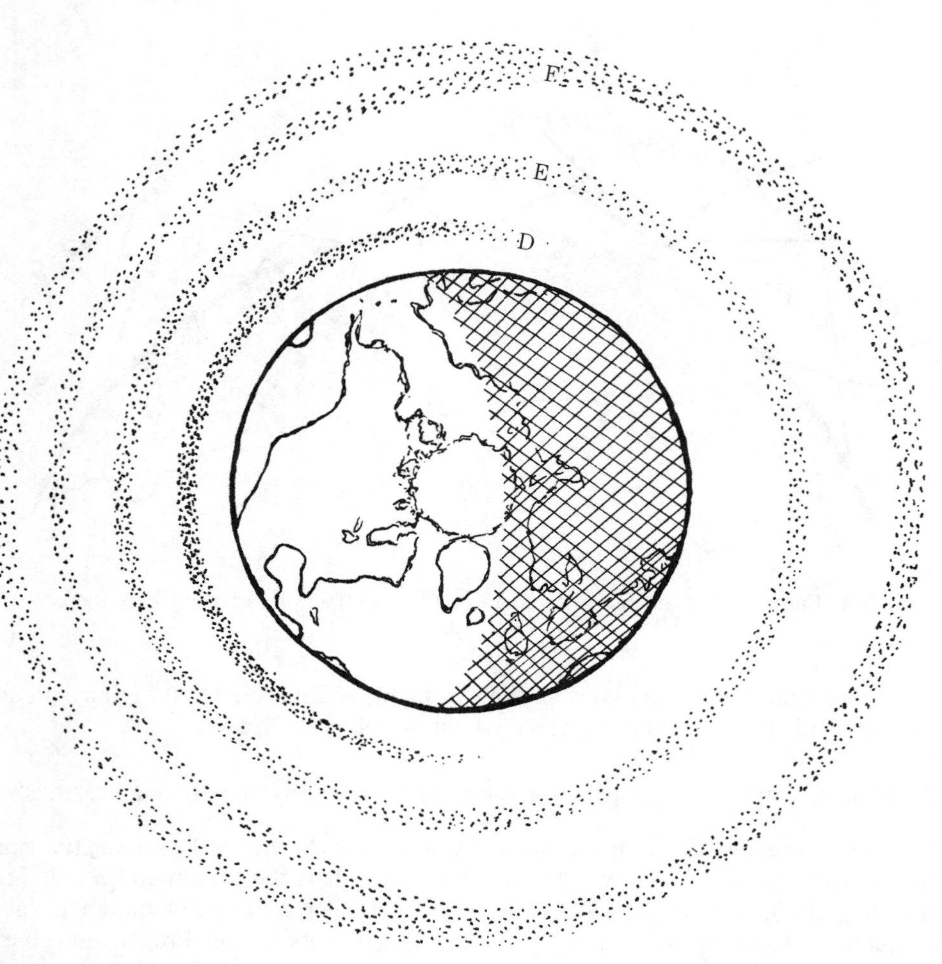

5-5 The ionosphere is much different on the daylight side of the earth than it is on the night-time side.

Q. Why can I hear some overseas stations, such as Deutsche Welle, The British Broadcasting Corporation, and Radio Japan so clearly, while other stations broadcasting from the same parts of the world have much weaker signals?

A. It's possible that they run higher transmitter power, or you could be hearing a broadcast from a relay station (Fig. 5-8). Several big international shortwave broadcasters, including the ones you just mentioned, use satellite technology to relay their programs to Canada or one of the islands in the Caribbean where they are re-broadcast on shortwave.

Q. What causes the electrical noise I sometimes hear on my shortwave radio?

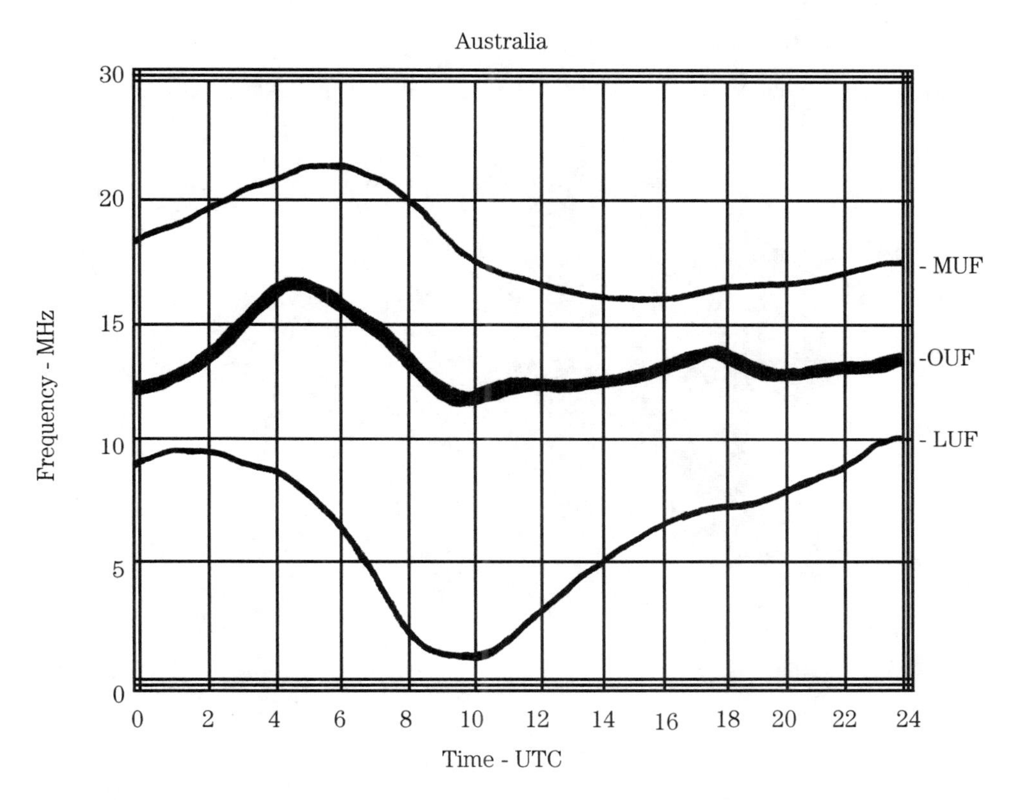

5-6 Propagation charts help you determine the best time and frequency for DXing.

A. Many things in your neighborhood can cause electrical noise. Thunderstorms, powerlines, computers, fluorescent lights, power tools, kitchen appliances, microwave ovens, and outdoor lights are only a few of the possible culprits (Fig. 5-9).

Solutions to noise problems can range from simple to extremely difficult. But your first task is to find out what is causing the problem!

If you have fluorescent lights in your house or apartment, turn them off and see if the noise is still there. If the noise disappears, they are the noise source and can be replaced with regular light fixtures. If you use a computer, it could be radiating noise to your shortwave radio. To solve that problem, you'll have to do your shortwave listening away from your computer.

Short periods of noise can be caused by any number of household appliances operating within a city block of you. It's hard to pin down the cause of that kind of noise if it's coming from outside your house but, thankfully, it usually doesn't last long enough to be a major disruption to your listening.

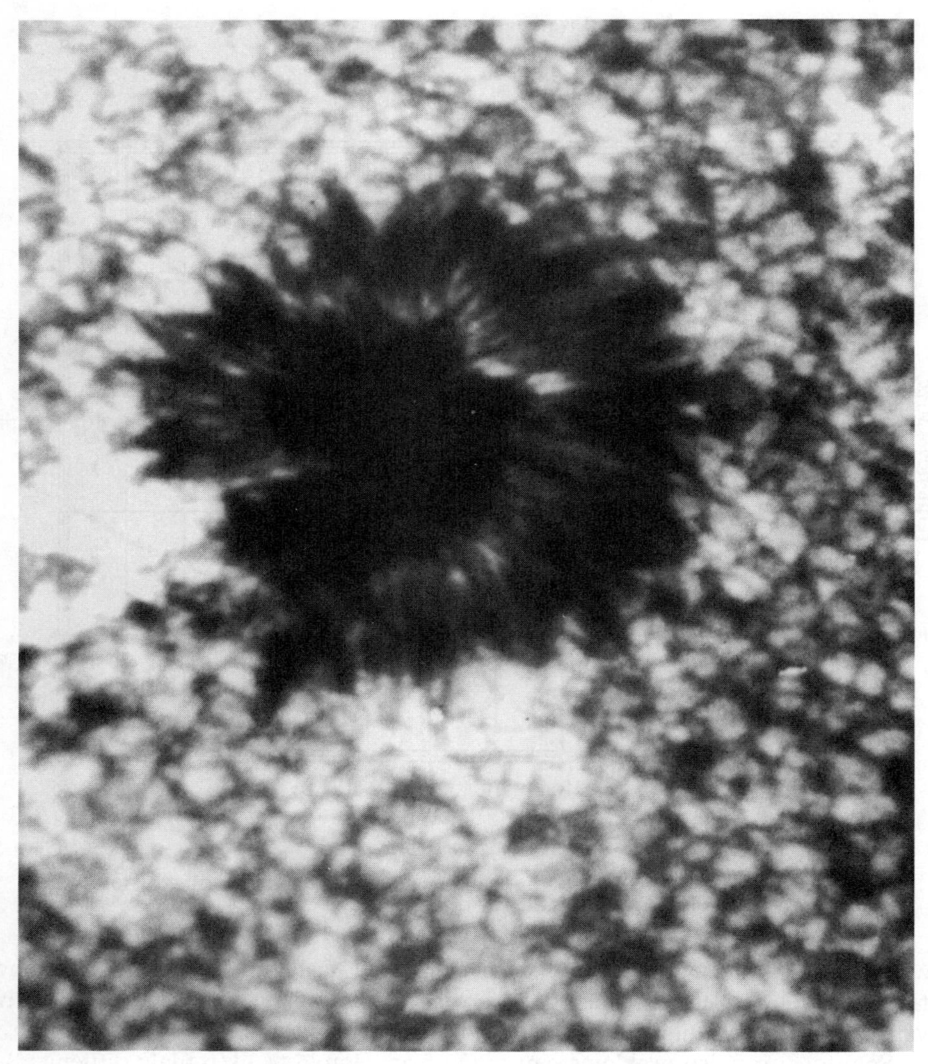

5-7 This sunspot, photographed by a NASA space probe, is roughly the size of the United States.

Noise that comes over your radio only at night could likely be caused by a nearby dusk-to-dawn security light. If you can locate the source, you can explain the problem and ask its owner to replace it.

If you hear a loud buzzing noise on your shortwave radio 24 hours a day, it could very well be coming in on the power line. Ask a representative from the power company to see if the noise is being caused by a problem in their equipment. If it is, they can probably repair it.

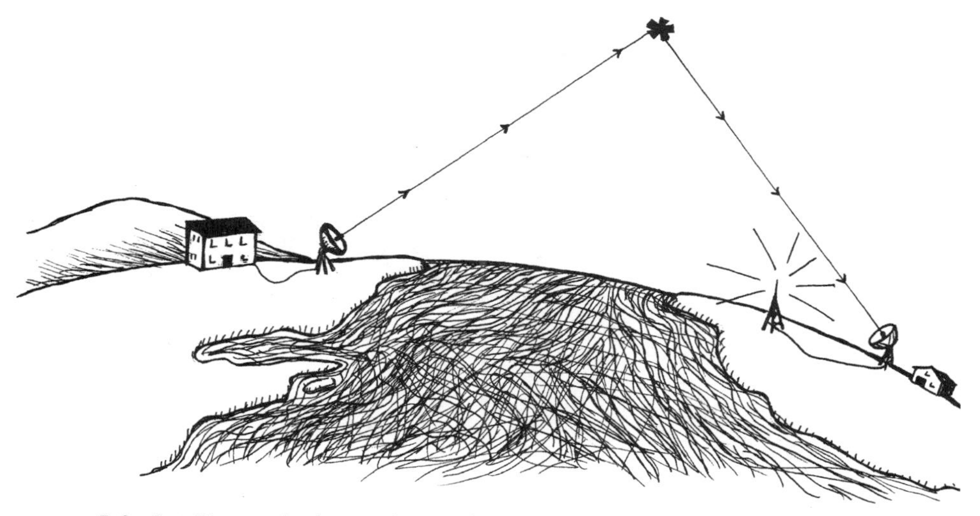

5-8 Satellites and relay stations make it easy to hear some foreign countries.

5-9 Many things can cause electrical interference.

5-10 Once a radio wave leaves earth, it travels forever in space.

Q. What happens to radio waves once they escape the ionosphere?

A. When radio waves pass through the Earth's ionosphere, they keep going forever in space. Every day, signals from thousands of television stations, radio stations, hams, police dispatchers, and countless other transmissions begin their endless journey through the universe (Fig. 5-10).

Someday, intelligent beings on distant worlds might hear them!

6
Selecting the right listening equipment

Q. What kind of a radio do I need to hear shortwave stations?

A. There are dozens of shortwave radios on the market that can pick up stations broadcasting from foreign countries. Each receiver has its advantages—as well as its drawbacks. The radio you should buy depends on your budget, listening requirements, and personal preference.

Q. How much do shortwave receivers cost?

A. If you're new to the hobby and don't want to invest a lot of money in a shortwave receiver, you can buy a "bargain basement," no-frills radio from $30 to about $100. These radios cover the AM and FM bands and most of the international shortwave bands. To keep prices low, these receivers rarely have digital frequency readout and they almost never provide single sideband reception. They are capable of receiving dozens of major shortwave broadcasters, such as the British Broadcasting Corporation, Radio Canada International, Radio Moscow, and Deutsche Welle, but they can't bring in many of the weaker stations you could hear on more sophisticated models.

Still, these low-priced shortwave sets are a fun and inexpensive way for a person on a budget to enter the exciting world of international radio listening. If you decide to invest in a better receiver later on, you can always use it for a back-up portable. The radios shown in Figs. 6-1 and 6-2 are typical of those sold in the $80 to $120 price range.

Q. What kind of reception can I expect on multi-band radios that cover shortwave, police, weather, air, and other services?

A. A number of multi-band radios on the market costing from $75 to $100 have one or two shortwave bands, as well as several bands marked "Police," "Air," "CB," "Weather," etc. Such radios are nearly always poorly designed, and are extremely

Grundig (Lextronix, Inc.)

6-1 The Grundig Traveler II.

Sangean America, Inc.

6-2 The Sangean SG 621.

difficult to tune accurately. If you already own one, you can use it to get a rough idea of what international listening is all about, but even a $30 bargain-basement shortwave radio usually will give you better performance on the international bands.

Q. What features should I look for in a medium-priced shortwave radio?

A. If you're investing much more than $120 in a portable shortwave radio, you should insist that it has digital frequency readout so you'll never have to guess which frequency you're on. Digital frequency readout makes it easy to find stations you want to hear and simplifies the process of identifying unfamiliar stations.

Another important feature to look for is single sideband (SSB) reception. Most shortwave receivers costing $175 or more have it. While most shortwave stations that broadcast to the general public use the same double-sideband mode that AM stations do, owning a radio with single sideband capability opens up a whole new world of international listening. (Single sideband transmissions sound garbled if you listen to them on a radio without a SSB control.)

Most ham radio operators use single-sideband transmitters, as do many utility stations. If you're interested in listening to pirate radio stations, you should definitely buy a radio with single sideband, since many North American pirates use that mode.

Many medium-priced receivers (as well as a few lower-priced models) have memory banks where you can store the frequencies of your favorite stations, and hear them at the push of a button.

Also, you should try to select a receiver that has continuous 3 to 30 MHz coverage through the shortwave bands. Several digital readout receivers on the market only cover portions of the shortwave frequency spectrum, making it impossible for you to tune in some stations you could otherwise enjoy.

The Sangean ATS 800 (Fig. 6-3) usually sells for around $80 to $100. It has digital frequency readout and 20 memories you can preset to your favorite stations. It receives shortwave, AM, FM, and FM stereo, and comes with headphones for private

6-3 The Sangean ATS 800.

listening. On the negative side, it only covers shortwave frequencies between 3.2–7.3 and 9.5–21.75 MHz, and doesn't provide single sideband reception.

The Panasonic RF-B65 receiver shown in Fig. 6-4 is typical of what you will find in the $175–$250 price range. It has SSB and continuous shortwave coverage, and lets you store 36 of your favorite station's frequencies in its memory banks.

6-4 The Panasonic RF-B65.

The Realistic DX-390 (Fig. 6-5), another nice radio in the same price range, has continuous shortwave coverage, SSB, and a 45-frequency memory bank. This receiver can be found at Radio Shack stores nationwide.

If you want to record what you hear, consider buying a Sangean ATS 818CS (Fig. 6-6). It has all the features mentioned above (including a memory bank that holds 45 frequencies), plus the ability to record and play standard sized cassette tapes. It normally sells for around $250.

On the upper end of the medium-price range, you can find high-tech radios such as the Grundig Satellite 700 (Fig. 6-7) for around $475 to $550. It comes preprogrammed with the main frequencies of 15 major international broadcast stations, provides you with 512 frequency memories, and can be upgraded to furnish you with 2,048 memories. The Grundig Satellite 700's data monitor display shows you the frequency, station name (of pre-programmed stations), signal strength, and the time. It has single sideband reception, a narrow/wide bandwidth selector, excellent sound quality, and many other features.

Q. I plan to do a lot of listening—do batteries last very long in shortwave radios?

Tandy Corporation/Radio Shack

6-5 Realistic DX-390.

Sangean America, Inc.

6-6 Sangean ATS 818CS.

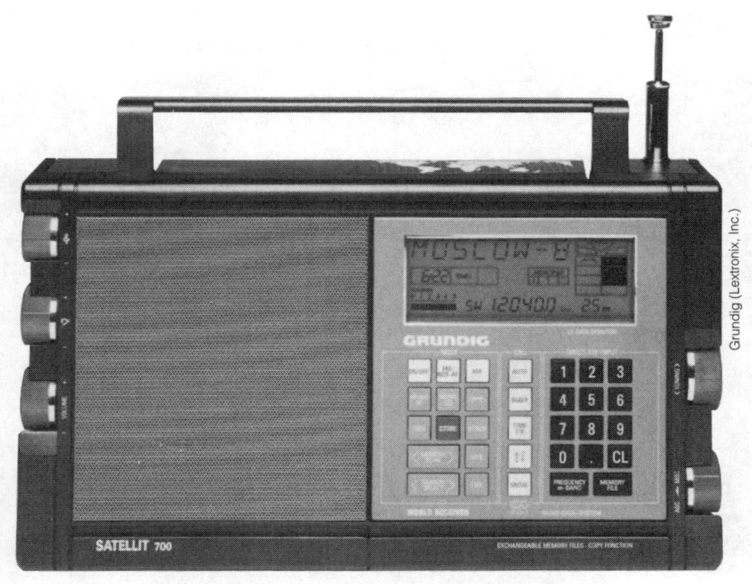

Grundig (Lextronix, Inc.)

6-7 Grundig Satellite 700.

A. They'll last about as long in a shortwave radio as they would in an AM/FM radio. However, if you prefer not to use batteries for at-home listening, most manufacturers give you the option of operating your receiver on house current by means of an adaptor.

If an adaptor didn't come with your radio, you can purchase one separately. Universal power adaptors such as the one shown in Fig. 6-8 can be adjusted to provide anywhere from 3 to 12 volts dc. They come with several sizes of plugs on the interconnect cord, so one should fit your radio's socket. When using such an adaptor, make sure that both the voltage and polarity are set correctly for your receiver, or you could damage your radio. If you are in doubt as to how any adjustment on your universal power adapter should be set, ask an electronics supply dealer to help you.

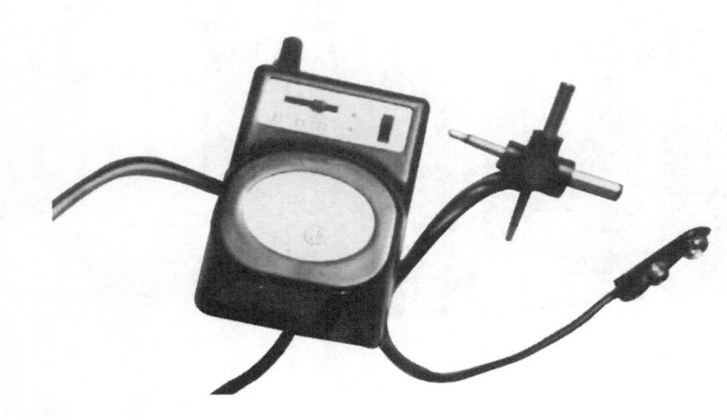

6-8 A universal power supply.

Q. What should I look for in more expensive shortwave receivers?

A. Shortwave radios in the high-price range have better sensitivity and selectivity than radios in the low- and mid-price range. Thus, they should be able to pick up weaker stations than less expensive models can and should have adequate filters to prevent powerful stations from coming in over weak stations on nearby frequencies.

The Lowe HF 150, priced at around $600, and the Lowe HF 225, priced at around $800 (Fig. 6-9), are high-quality radios that are very easy for even a beginner to operate. These British-manufactured receivers give you full shortwave frequency coverage, single sideband reception, memory banks, and selectable filters that permit you to hear weak stations when they're sandwiched between stronger ones. In North America, these award-winning Lowe receivers can be purchased through Electronics Equipment Bank and Universal Radio.

Lowe Electronics, Ltd.

6-9 The Lowe HF-150 and HF-225.

The Icom IC-R1 (Fig. 6-10) hand-held receiver lets you tune in stations between 2 to 905 MHz. This means you can receive AM, FM, shortwave, and VHF/UHF frequencies—all on the same radio! This pocket-sized portable stores up to 100 frequencies in its memory bank, and can be programmed to scan through groups of frequencies until it locates a station. It has a built-in rechargeable lithium battery and a 24 hour clock that can be programmed to turn the radio on and off whenever you like. The IC-R1 is priced at around $500.

6-10 Icom IC-R1 scanning receiver.

Table-top communications receivers, such as the Drake R8 Communications Receiver (Fig. 6-11) and the Icom IC-R72 (Fig. 6-12) are a great investment for shortwave listeners who have had some experience with their hobby and want a receiver that brings in the kind of rare catches that average receivers miss. Both the Drake R8 and the Icom IC-R72 have numerous controls and adjustments that are desired by

R. L. Drake Company

6-11 Drake R8 Communications Receiver.

Printed with permission. Icom is a registered trademark of Icom, Inc.

6-12 The Icom IC-R72 Communications Receiver.

experienced listeners—but could be confusing if you're new to the hobby. They're priced in the $800 to $1000 range—but considering the many years of listening enjoyment they can give you, it's a real value!

Q. Where can I find reviews of shortwave receivers?

A. New shortwave receivers are reviewed in publications such as the *World Radio-TV Handbook*, *Passport to World Band Radio*, *Monitoring Times*, *Popular Communications*, and a number of DX club bulletins. They are also reviewed on DX programs such as Radio Nederland's "Media Network " and HCJB's "DX Party Line."

If you'd rather save money and buy a used or older model shortwave receiver, some good places to look for information are: *The Radio Hobbyist's Buyers Bluebook*—by Bob Grove (Fig. 6-13), *Shortwave Receivers Past and Present*—by Fred Osterman, *The WRTV Equipment Buyer's Guide*—by W. Bos and J. Marks, *Buying a Used Shortwave Radio*—by Fred Osterman, and *Popular Communications' Communications Guide*. You can also check the library for back copies of *Passport to World Band Radio* or *The World Radio-TV Handbook*. If you'd like to buy a used

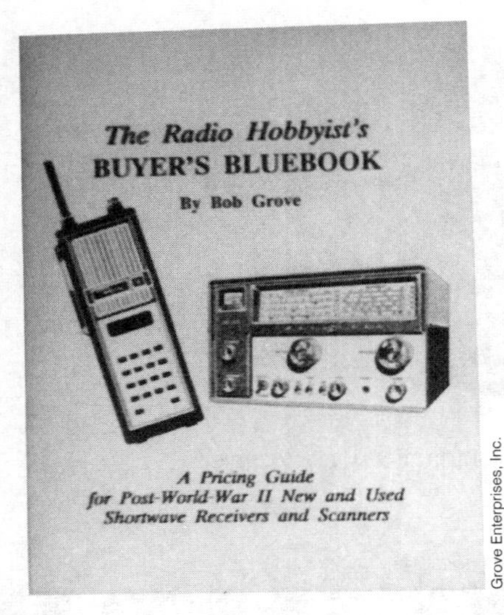

6-13 *The Radio Hobbyist's Buyer's Bluebook*, written by Bob Grove, is an excellent source of information on used shortwave receivers and scanners.

shortwave receiver but don't know where to find one, the local ham radio club is a good place to start your search.

Q. Are there shortwave radios available for my car?

A. Yes! If you think you'll be doing a lot of your shortwave listening in the car, you should look at the Philips DC-777 receiver (Fig. 6-14), priced at around $350. It features a cassette player, AM, FM/FM stereo, shortwave, and a memory bank that holds 20 frequencies. The Philips DC-777 is available through most shortwave mail-order dealers.

Q. Are shortwave radios available in kit form?

6-14 Philips DC 777 AM/FM/Shortwave radio and cassette deck.

A. Yes they are! If you're a beginner, or you want to buy a gift for a child or teenager with an interest in shortwave radio or electronics, the Radio Shack AM/Shortwave Radio kit (Fig. 6-15) is a nice way to get started in a hobby that can bring a lifetime of enjoyment. It requires no soldering, costs under $15, and brings in a number of AM and shortwave broadcast stations.

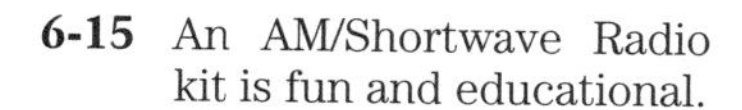
6-15 An AM/Shortwave Radio kit is fun and educational.

If you're more experienced and don't mind doing some soldering, the MFJ-8100 World Band Radio kit (Fig. 6-16) might be just what you're looking for. It covers the main shortwave and ham radio bands, and has single sideband reception. You can order the MFJ-8100 from MFJ Enterprises, Inc., (1-800-647-1800) for around $60 plus postage and handling. Or, you can buy one wired and tested for around $80 plus p & h.

6-16 MFJ World Band Shortwave Radio kit.

Q. What kind of antennas are available for shortwave listening?

A. If you're a beginner, the built-in extendable "whip" antenna is all you'll need for a while, but sooner or later, you'll get to the point where you want to hear more stations and decide to put up an external antenna.

One of the easiest to use and least expensive shortwave antennas is the Sangean ANT-60 (Fig. 6-17). It extends to 23 feet and can be rewound into its pocket-size case when not in use. Radio Shack carries a similar antenna in their stores nationwide. This type of antenna sells for $9–$15.

Longwire antennas (Fig. 6-18) and dipole antennas are also easy for beginners to

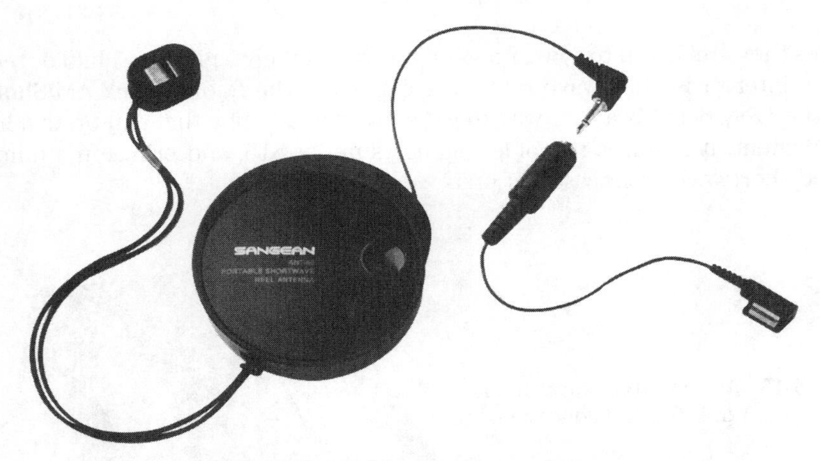

6-17 Sangean ANT-60 portable shortwave antenna. Sangean America, Inc.

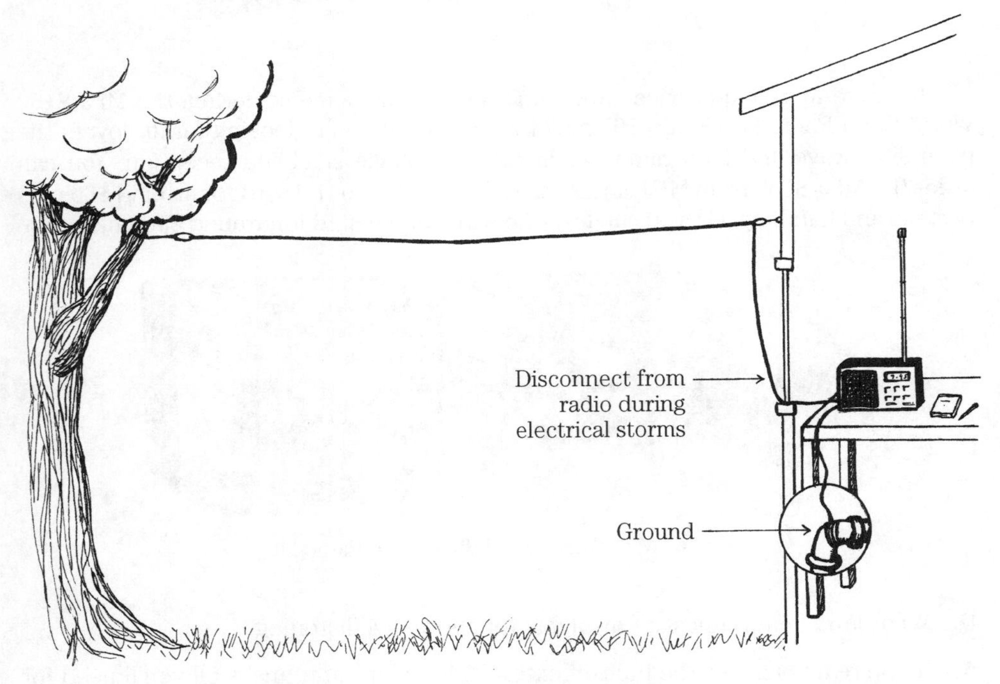

6-18 Long wire antennas are easy to erect.

set up. Radio Shack sells a 75-foot shortwave longwire antenna kit for under $10. They also sell a 65 foot "trap dipole" antenna especially designed to improve your reception of stations on the shortwave bands for around $40. Many shortwave mail-order companies offer a variety of antennas for international broadcast listening at reasonable prices.

For safety and to prevent electrical noise interference, outdoor antennas should always be kept as far away from power lines as possible.

Q. What is a tuner or preselector?

A. Preselectors amplify the station you want to hear, while rejecting electrical noise and other unwanted signals that don't belong on that frequency band. A tuner is a passive device that matches the impedance of the radio to the impedance of the antenna. They are connected between your receiver and your external antenna. The MFJ LW/MW/SW Preselector/Tuner (Fig. 6-19) can improve reception on the AM as well as the shortwave bands, and is priced at around $40.

6-19 A MFJ SW/MW/LW antenna tuner.

MFJ Enterprises, Inc.

Q. What is an active antenna?

A. An active antenna is a device that electronically boosts radio signals before they reach your receiver. They make weak signals stronger and allow you to hear stations you've never picked up before.

The Radio Shack Amplified Shortwave Antenna (Fig. 6-20) costs around $30. It is one of the least expensive active shortwave antennas available, and can boost the signal strength of any station. The Radio Shack Amplified Shortwave Antenna does a nice job of boosting signals using only its built-in whip antenna, or it can give you even better performance if you connect it to an external antenna.

Figure 6-21 shows two MFJ Enterprises active antennas. Fig. 6-21 (A) is a MFJ Active Antenna LF/HF/VHF—Model MFJ 1022—very easy to use. It improves reception on the AM, shortwave, and scanner frequencies up to 200 MHz. The MFJ 1022 sells for under $50. Figure 6-21 (B) is a MFJ-1020A Indoor Active Shortwave Antenna. It improves reception and reduces noise from 0.3 to 30 MHz. The MFJ-1020A is priced at around $80. Both models can be connected to an external antenna.

The MFJ Remote Active Antenna—Model MFJ 1024 (Fig. 6-22) amplifies signals in the AM and shortwave bands. The 54" outdoor antenna can be mounted outside your house or apartment window and comes with 50 feet of cable that connects it to the control box section, which sits next to your receiver. The MFJ 1024 sells for around $130.

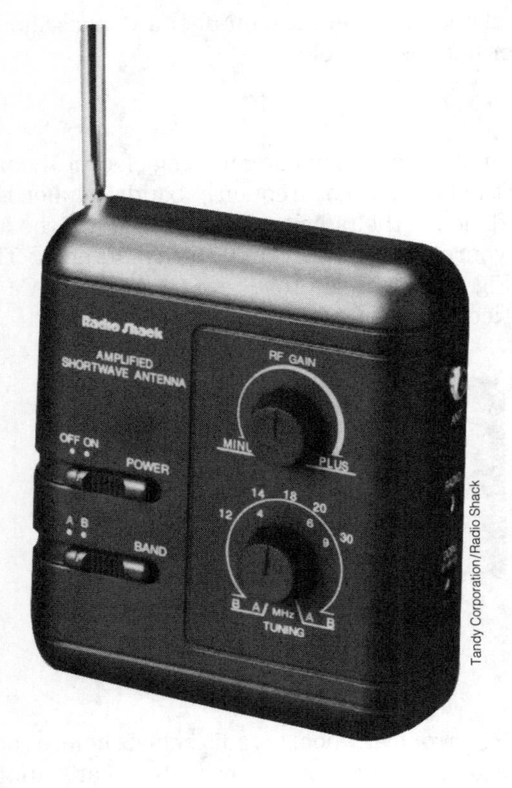

6-20 A Radio Shack Amplified Shortwave Antenna is an inexpensive way to improve your receiver's performance.

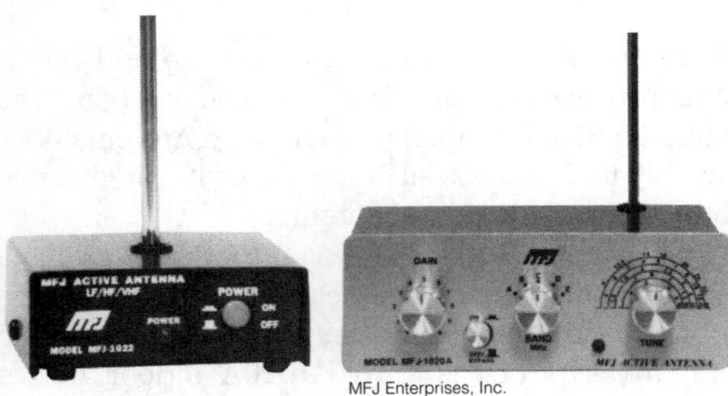

6-21 These MFJ active antennas boost signals before they reach your receiver.

Q. What kind of radio do I need to hear stations broadcasting on frequencies above 30 MHz?

A. Scanners are commonly used to hear stations broadcasting on the VHF (Very High Frequency) and UHF (Ultra High Frequency) bands. Scanners operate by "scanning" groups of pre-programmed frequencies for a transmission. When they find a transmission, they stop scanning, wait until the broadcast is over, then start scanning again. Most scanners cover frequencies ranging from 30 to 400 MHz—and some newer models can cover frequencies from 25 to 950 MHz or above.

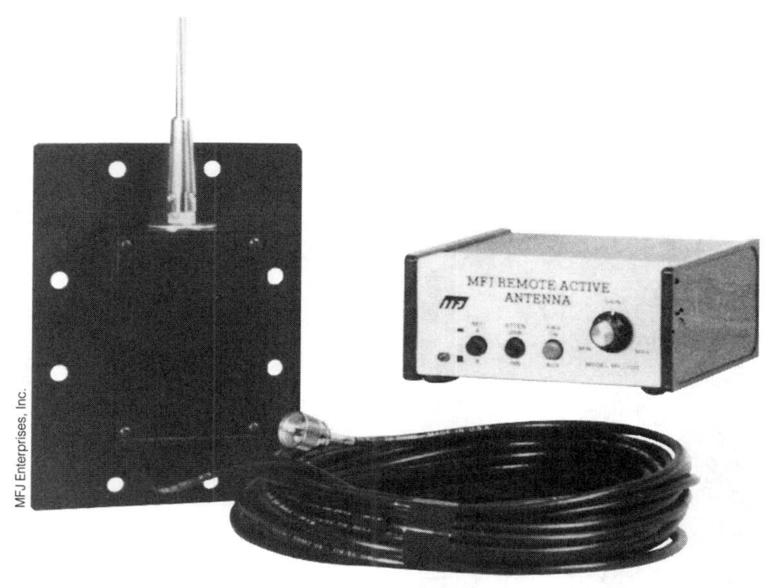

6-22 A MFJ Remote Active Antenna can help pull in stations you might otherwise miss.

Many people think of scanners as "police radios," but law enforcement communication is only a small part of what you can hear. Depending on which frequencies you decide to program into your scanner, you can hear aviation communications, ships, ham radio operators, truck drivers, utility companies, business radio, and many other kinds of transmissions.

Scanners come in a wide variety of styles. Basically, they are (1) hand-held (Fig. 6-23), (2) base and (3) car scanners. Some scanners, such as the Uniden BC350A (Fig. 6-24), come pre-programmed to scan through hundreds of frequencies for police, weather, air, marine, and emergency communications, also has 50 user-programmable memories to store frequencies of other services you would like to monitor. The Uniden BC350A can be used at home as a "base" scanner, or it can be mounted in your car (if you live in a state where using a scanner in your car is legal). Many mail-order companies sell it for under $150.

Q. How far away can you hear VHF/UHF transmissions?

A. The range you can expect to get from your scanner depends on many factors: your location (do you live in a valley or high on a hill?), the size of city or town you live in (scanner listeners in large cities have trouble hearing distant transmissions because of the large number of local frequency users), and other factors, such as electrical interference caused by computers, flourescent lights, and other devices.

The power of the transmitter also has a lot to do with how far away the signal can be heard. Police dispatchers, fire departments, utility companies, and other services that must communicate over several miles or more use much higher power levels than walkie-talkie-type business radios intended for communicating with people a

Tandy Corporation/Radio Shack

6-23 The Realistic Pro-43 Hyperscan scanner has a 200-frequency memory.

mile or less away. Cordless telephones, one of the lowest-powered transmitting devices to operate on scanner frequencies (46.610–46.970 MHz), can normally be heard only a few blocks away.

However, unusual weather patterns or ionospheric conditions can sometimes create "skip" conditions that transport scanner band transmissions to listeners hundreds of miles away!

Because an airplane's transmission antenna is mounted at such a high altitude, communication from pilots to ground control can be heard at great distances. Com-

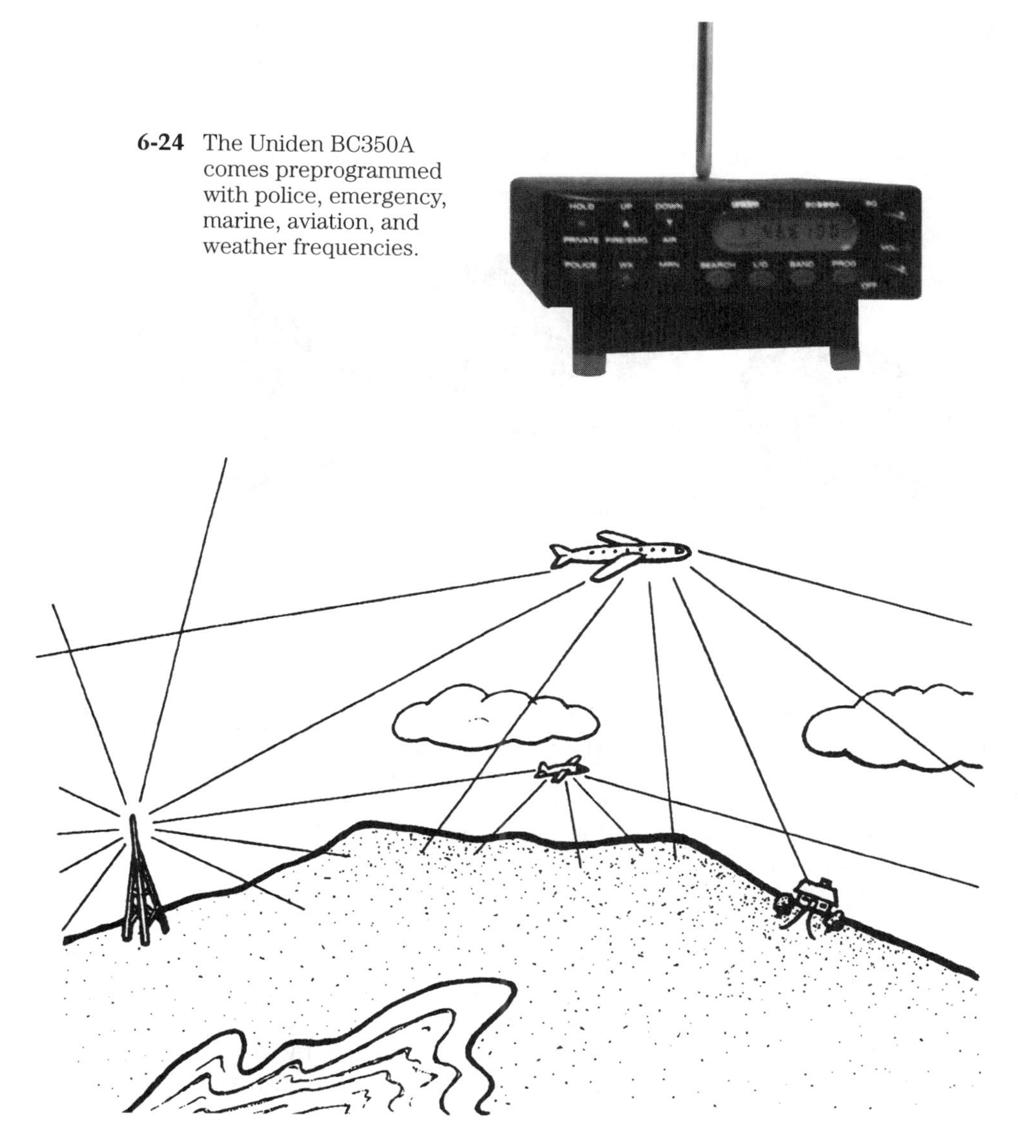

6-24 The Uniden BC350A comes preprogrammed with police, emergency, marine, aviation, and weather frequencies.

6-25 Transmissions from high-flying aircraft are heard hundreds of miles away.

mercial, private, and military jet aircraft can be heard over several states when flying at their cruising altitudes of 20,000 to 30,000 feet or higher (Fig. 6-25).

Q. What kinds of antennas are available for scanners, and how much do they improve your listening range?

A. The built-in "whip" or rubber-coated antenna that comes with your scanner will provide you with good local coverage under most conditions. If you want to extend your scanner's range, numerous designs of external antennas are available. Some are made for general scanner band listening, and others are designed to greatly improve reception on certain specified sections of the scanner frequency spectrum.

6-26 Grove's Indoor Scanner Antenna System can greatly increase your scanner's reception range. Grove Enterprises, Inc.

If you live in an apartment, Grove's Indoor Scanner Antenna System (Fig. 6-26) might be exactly what you need. The 66-inch antenna can easily be hidden around the edge of your window, and the PRE 4 amplifier box boosts transmissions on frequencies from 30 to over 1000 MHz. The entire system costs around $100.

The Discone Scanner/Ham Roof Antenna (Fig. 6-27), available from Radio Shack, improves your reception on scanner frequencies between 25 and 1300 MHz. If you decide to become a ham, you can also use it as a transmit antenna on the 144, 220, 440, and 1296 MHz amateur radio bands. It is priced at around $60.

Electronics stores and mail-order companies that specialize in shortwave and scanning equipment have many types of scanner antennas for you to select from. Mobile scanner antennas for your car are available almost everywhere scanners are sold. And if you want to dramatically improve the scanning range of your base scanner, look into a directional "beam" antenna that mounts on a rotor and can be pointed at the area you want to monitor.

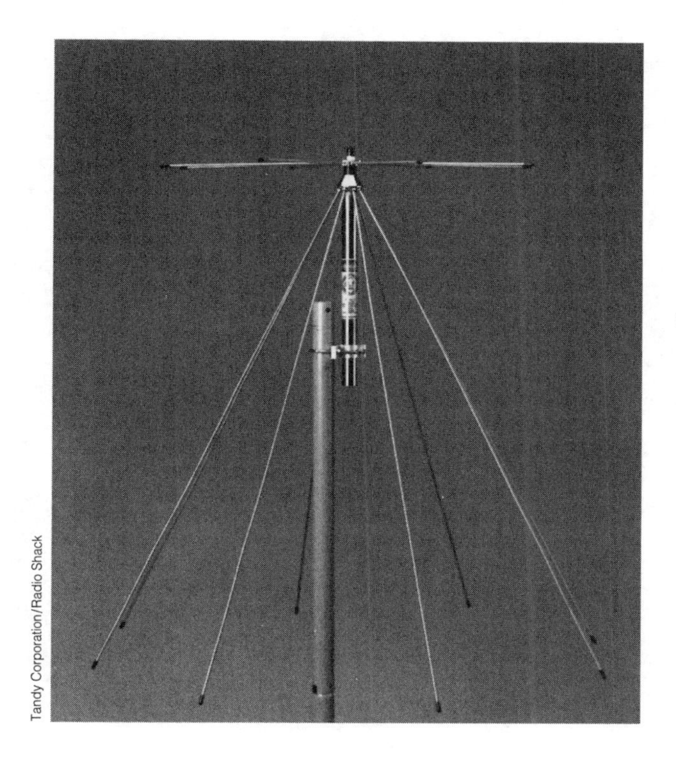

Tandy Corporation/Radio Shack

6-27 The Discone Scanner/Ham Roof Antenna improves your reception on frequencies between 25 and 1,300 MHz.

7
Keeping up with the latest information

Q. How can I keep up-to-date with the latest in shortwave?

A. As a shortwave listener, you have many ways of keeping up with the latest information. Schedule changes, receiver and antenna reviews, as well as information on new stations and technical aspects of the hobby, are available through books, magazines, radio clubs, and shortwave DX programs.

Q. How current are the *World Radio-TV Handbook* and *Passport to World Band Radio*?

A. Both books are published annually, and last-minute schedule changes are made as close to the publication date as possible. Even though some broadcast times and frequencies may change before the year is out, most other information, such as station addresses, transmitter power, QSL policies, and product reviews, are extremely useful for some time to come.

Q. Where can I order shortwave books?

A. Your local bookstore can order The *World Radio-TV Handbook*, *Passport to World Band Radio*, and books published by well-known companies, such as TAB /McGraw-Hill and the American Radio Relay League. You can also obtain these, as well as many of the other books listed in Appendix D, through shortwave mail-order companies. Their catalogs (Fig. 7-1) are filled with the latest shortwave books, receivers, antennas, etc. Appendix C lists the addresses of shortwave mail-order outlets throughout North America.

Q. Which magazines contain information on shortwave schedule changes?

7-1 Shortwave mail-order companies have many excellent books, receivers, antennas, and other related products for you to select from.

A. *Monitoring Times* (Fig. 7-2) and *Popular Communications* are the largest circulation shortwave magazines in North America. They provide detailed schedule information on dozens of foreign broadcasters, and both include listings of shortwave programs broadcast in English.

Q. Can I obtain schedules directly from the stations I'm interested in hearing?

A. Most international shortwave broadcast stations keep a mailing list of listeners who wish to receive program schedules. These schedules, which are issued on a monthly, quarterly, or bi-annual basis, tell you when and where you can hear broadcasts to your part of the world. Many of these schedules are little more than a listing of times, frequencies, and languages, but some stations, such as Radio Budapest International (Fig. 7-3), make their schedules into mini-magazines that provide detailed information on upcoming programs.

Q. Do shortwave stations provide their listeners with information on other international broadcasters?

A. Yes. Most international shortwave broadcasters have some sort of DX program. They know their listeners are hungry for up-to-date information about the hobby, and are eager to provide it. Some of the most popular DX programs are Radio Nederland's "Media Network", HCJB's "DX Party Line", WWCR's "Spectrum", and Glenn Hauser's "World of Radio", which can be heard on several North American shortwave stations.

Q. Where can I find schedules for shortwave DX programs?

A. *Monitoring Times* regularly publishes schedules of shortwave DX programs. You can also find information about DX shows in many radio club bulletins.

7-2 *Monitoring Times* is one of the most popular radio magazines in America.

Q. Do general electronics/technology magazines and newsletters publish information about shortwave radio?

A. Yes, some do. *Popular Electronics*, *Electronics Now*, and *NuTechnology Newsletter* (Fig. 7-4) occasionally publish information on shortwave broadcasting and antennas, or review new shortwave receivers. With the increasing popularity of shortwave listening, more publications are likely to make information on the international radio listening scene available to their readers.

BUDAPEST
INTERNATIONAL
No. 2. 1993

BUDAPEST CHARITY BALL IN JUNE

7-3 Radio Budapest International publishes a mini-magazine for their listeners.

Q. What is the best way to select a DX club?

A. You can obtain a sample copy of most DX club bulletins by sending a dollar or two to cover printing and postage costs. Some clubs report on shortwave station news only, while others cover the AM band, shortwave utility stations, scanner news, ham radio, etc. Most new hobbyists order a sample bulletin from several different clubs and look them over before deciding which one(s) they like best.

The North American Shortwave Association (NASWA) is one of the most popular clubs in the USA. Their monthly bulletin (Fig. 7-5) is filled with information on shortwave programs, QSL collecting, and book and product reviews.

Q. Do any shortwave clubs specialize in DXing pirate stations?

A. Yes! The Association of Clandestine radio Enthusiasts (A*C*E) monthly bulletin (Fig. 7-6) is filled with up-to-date information on pirate and clandestine broadcasts.

If you're really interested in pirate stations, you'll probably want to get a copy of *The Pirate Radio Directory*, by George Zeller and *The Worldwide Pirate Radio Logbook* by Andrew Yoder (Fig. 7-7). Both books are published annually.

The Pirate Radio Directory contains a brief write-up about all the North American shortwave pirate stations that have been active in the past year. It describes the kind of programming each station airs, the frequency it normally uses, and how active it has been throughout the year. *The Worldwide Pirate Radio Logbook* lists the time, date, frequency, and station name for almost every shortwave pirate broadcast that took place in North America and Europe during the past year. It also includes international pirate maildrop addresses.

TO SUBSCRIBE: 26 issues of *NuTechnology Newsletter*, for U.S, $30; for Canadian subscribers: $40., U.S. funds on a U.S. bank; for overseas subscribers: $65. U.S. funds on U.S. bank. Mailed First Class every other Saturday.

NuTechnology

New Products and Applications

Published every other Friday by Hart Publishing, 767 S. Xenon Ct., #117, Lakewood, CO 80228

CONTENTS

VOL. 1 NO. 4
AUGUST 6, 1993

New telecom network gives public on-line access to national research database

A telecommunications network that gives the public on-line access through public libraries to the rapidly expanding Internet was unveiled August 4, by Bellcore, New Jersey Bell, and M.A.I.N. — a consortium of 32 Morris County, New Jersey public and academic libraries.

The Internet — a "network of networks" — is a worldwide patchwork of more than 11,000 computer networks used mainly by academic, government and commercial researchers. An increasing number of schools and individuals are becoming part of the Internet community.

The new network, MORENET (Morris Research and Education Network), lets patrons of Morris County's 32 public and academic libraries — including elementary and secondary grade students — bridge into the Internet and directly access some of the nation's most massive information databases. Included among those are ones operated by the Library of Congress and National Aeronautics and Space Administration (NASA).

"Our goal is to open up information opportunities to students and the general public, regardless of their individual economic circumstances," said Judith Weinstein, president of the Morris (County) Automated Information Network (M.A.I.N.) planning council.

"With MORENET, the residents of Morris County, NJ will join professionals and researchers in having access to — from publicly provided computers — the information explosion brought about by electronic internetworking."

Students, for example, can locate and retrieve works of literature, examine holdings of thousands of libraries, and share current news events with students nationwide.

While other U.S. communities have established library information networks, they are typically local in nature. Internet access is sometimes provided by universities, but rarely by municipal libraries. In addition, MORENET is the first to offer use of the Bellcore-developed high-speed text browsing and retrieval software system called Superbook—one of the most effective systems for location of specific information in documents thousands of pages in length.

MORENET is also a unique collaboration between industry and local government to enhance public access to the larger world of electronic information. "We believe that if the public can easily access the Internet, they will be encouraged to take advantage of the variety of services and information-sharing capabilities currently available to researchers and the government," said New Jersey Bell President Al Koeppe.

For example, said Judy Weinstein, "The ability to plug into the Internet could dramatically change the way librarians support the public's information needs. With networks like MORENET, librarians can offer people a wider range of up-to-date reading and reference materials just by typing in a few simple commands."

With MORENET, Morris County library cardholders and librarians will have easy and convenient access to a growing number of Internet information resources. These include: - Library of Congress catalogue - NASA Spacelink (related to space technology and news) - Oceanic (oceanography information) - University of California Libraries catalog - CARL (Colorado Academic Research Libraries) catalog (includes databases from at least 13 education and governmental libraries) -

Continued on page 2

7-4 *NuTechnology Newsletter* covers developments in all fields of technology, including shortwave and ham radio. Hart Publishing

Q. What kind of current information is available for people interested in decoding RTTY transmissions?

A. Universal Radio publishes *The RTTY Listener* (Fig. 7-8), containing information on RTTY and other digital modes of transmission, and reviews of new RTTY decoding products. *Monitoring Times*, *Popular Communications*, and a few DX clubs also publish information on how and where to intercept RTTY transmissions.

Q. Which magazines and clubs report on offshore radio stations?

A. *Monitoring Times*, *Popular Communications*, and many DX clubs report on offshore radio stations, especially if they broadcast on the shortwave bands.

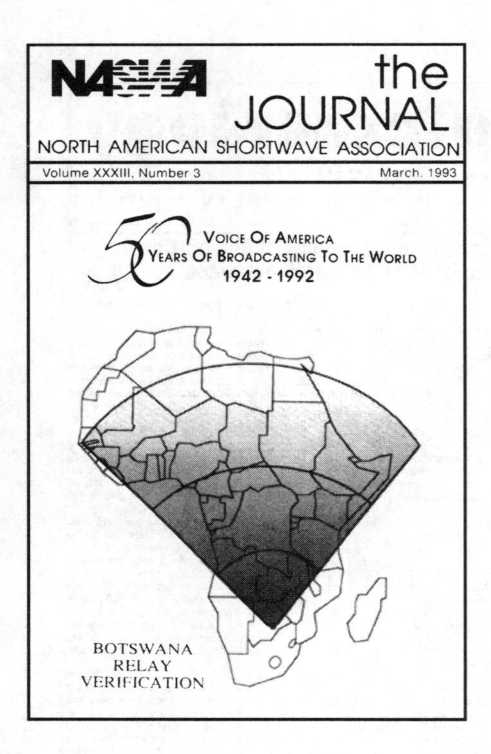

7-5 The North American Shortwave Association publishers a monthly bulletin, *The Journal*.
NASWA

7-6 The A*C*E (Association of Clandestine radio Enthusiasts) bulletin is devoted entirely to reporting on pirate and clandestine radio activity.

7-7 *The Pirate Radio Directory* and *The Worldwide Pirate Radio Logbook* are good sources of information on pirate radio broadcasting.

Several European magazines and clubs specialize in reporting on offshore and pirate radio. *Offshore Echo's* magazine (Fig. 7-9) is filled with articles on offshore stations, past and present. *FRS Goes DX* (Fig. 7-10) covers the latest developments in both pirate and offshore broadcasting activity.

Q. Are any computer bulletin boards available for shortwave hobbyists?

A. Yes. Some DX clubs and a number of shortwave radio enthusiasts run bulletin boards where computer and modem-equipped shortwave listeners can exchange information. You can find out how to contact them through DX club bulletins, shortwave magazines, and international DX programs.

Q. Does the shortwave hobby have conventions where people can go to meet other shortwave listeners?

A. *Monitoring Times* and *Popular Communications* magazines (Fig. 7-11) host annual conventions for shortwave listeners, featuring speakers who are experts in their field of DXing. Shortwave radio dealers from all over the country come to promote their latest products, and international broadcasting personalities often make appearances, as well.

Most national, regional, and local shortwave listening clubs also have regular get-togethers for their members. If you don't live close to a shortwave DX club or can't attend the conventions, ham radio clubs are another good place to look for people involved with shortwave radio listening. Nearly everyone has a ham club nearby, and you'll probably find some members that share your interest in tuning foreign broadcasters.

The *RTTY Listener*

Issue 31 January 1993

The Ultimate DX *by Fred Osterman*

There are two major types of telescopes. Optical telescopes use lenses and mirrors to concentrate light. Radio telescopes focus radio waves. The basic operating principles are identical except that radio telescopes operate at wavelengths a million times longer than the light wavelengths of optical telescopes. Therefore they look very different.

One of the oldest and largest radio telescopes in existence is located just north of Columbus. The Kraus 110 meter space telescope is operated by the North American Astrophysical Observatory and Ohio State University. This huge instrument (the size of a football field!) is well suited for deep space research and mapping. In recent years the telescope has been dedicated to SETI (Search for Extra Terrestrial Intelligence). Specifically, the sky survey is presently searching the 1.4 to 1.7 GHz band for signals of extraterrestrial *technological* origin (intra-galactic RTTY?). This impressive, and highly automated, search was shown to the public during a recent open house. It was interesting to note the use of a computer controlled Icom R-7000, purchased from Universal Radio, in use, as part of a much more complex multi-channel system.

Operating and maintaining the Kraus 110 meter telescope is an enormous task. This is done by a nonprofit, volunteer organization with very limited institutional support. A monthly newsletter, *Signals*, is available to those contributing $25 or more. Inquiries and donations should be addressed to:

Dr. Robert Dixon, Director
OSU Radio Observatory
805 Dreese Labs
2015 Neil Ave.
Columbus, OH 43210

Press FAX is Back

Victor Nowik reports the popular FAXpress eminating from Buenos Aires can now be found on:

8167 kHz	60/288	2300 UTC
9241 kHz	60/288	2300 UTC
11480 kHz	60/576	2100 UTC

Printing 240 LPM FAX

RTTY Listener reader Ed Collins of Ohio is one of the fortunate few equipped to copy GOES FAX *directly* via the 1691 MHz down link. Ed's problem is printing the 240 LPM image on a dot matrix printer. Many printers can't "keep up". If you are *successfully* printing 240 LPM FAX to a dot matrix printer, please drop a note to Universal Radio (Attn: F. Osterman), indicating model of printer used.

Co-Editors

Robert E. Evans and Eric Sillick of Ontario, Canada prepared pages 3 through 13 of ***RL# 31***. As always, we appreciate their generous efforts.

VGA Monitors and R.F.I. *by Hugh A. Roberts*

I'm using an Icom R-9000 and a Toshiba SVGA monitor. The monitor has selectable video inputs, one can use a video cable with a DB-15 connectors or one with separate BNC inputs for the R, G, B, B&W and SYNC lines of the VGA signal. When the M-8000 arrived, I did not give much thought to the type of video cable to be used. I purchased a "VGA standard" cable with DB-15 connectors on both ends, hooked everything up and turned it on.

I noticed there was a lot of RFI (interference) with the M-8000 turned on between 8 and 14 MHz, sometimes enough to block out strong signals. With the M-8000 off there was no RFI. I rerouted my incoming antenna coax to stay away from the M-8000 and the monitor, resulting in a slight reduction in interference.

I noted that if the audio output was unplugged from the R-9000, the RFI decreased. I replaced the RCA plug type connecting cable with a shielded audio cable with shielded metal plugs and gained a five per cent decrease in RFI.

With the system running and the VGA cable unplugged, the RFI decreased about 95 per cent, leading me to conclude the cable was the culprit. I replaced the cheap wire cable with an SVGA cable containing coax conductors for each of the RGB, B&W and SYNC signals and obtained a significant decrease in RFI. I ordered an F240-77 toroid core from Palomar Engineers and was able to get five wraps of the video cable through it; another incremental decrease in RFI. I grounded the monitor and the remaining small amount of RFI disappeared completely!

C Band Update *by Mark Burkhart*

ANIK E1	Transponder: 10	Reuters press FAX
MORELOS A	Transponder: 10	VFT 50/A/03 and 10 Spanish Telex tfc.
MORELOS A	Transponder: 16	VFT 50/A/04, 08 and 16 Spanish Telex tfc.

7-8 *The RTTY Listener*, published by Universal Radio, Inc., keeps RTTY enthusiasts up-to-date on new stations and frequency changes. Universal Radio, Inc.

7-9 *Offshore Echo's* magazine is about offshore radio stations, past and present.

7-10 *FRS Goes DX* reports on pirate radio activity in Europe and around the world.

Popular Communications

7-11 Bob German speaks to shortwave enthusiasts at the 1993 Popular Communications SWL Convention in Virginia Beach, VA.

Appendix A
Shortwave station addresses

Here are the addresses of some of the more commonly heard shortwave radio stations. Check with the post office for proper overseas airmail postage rates.

Radio Tirana
International Service
Rrug Ismail Quemali
ALBANIA

Radiodiffusion Argentina Al Exterior—RAE
C.C. 555 Correo Central
1000 Buenos Aires
ARGENTINA

Radio Australia—ABC
PO Box 755
Glen Waverly, VIC 3105
AUSTRALIA

Radio Austria International
A-1136
Vienna
AUSTRIA

Belgische Radio En Televisie
PO Box 26
B-1000 Brussels
BELGIUM

Radio Nacional
C/P/ 04/0340
70323 Brasilia
BRAZIL

Radio Sofia
4 Dragan Tsankov Blvd.
1040 Sofia
BULGARIA

Radio Canada International
PO Box 6000
Montreal, PQ H3C 3A8
CANADA

CFRX/CFRB
Ontario DX Association
PO Box 161, Station A
Willowdale, Ontario M2N 5S8
CANADA

Radio CHU (Time & frequency)
National Research Council
Ottawa, ON K1A OR6
CANADA

China Radio International
Beijing 100866
PEOPLE'S REPUBLIC OF CHINA

Voice of Free China
P.O. Box 24-38
Taipei, Taiwan
REPUBLIC OF CHINA

Radio for Peace International
University for Peace
Apartado 88
Santa Ana
COSTA RICA

Croatian Radio
Studio Zagreb
PO Box 1000
41000 Zagreb
CROATIA

Radio Habana Cuba
PO Box 7026
Havana
CUBA

Czech Radio
12099 Parah 2
Vinohradska 12
Prague
CZECH REPUBLIC

Danmarks Radio
Radiohuset
DK-1999 Fredriksberg C
DENMARK

HCJB, Voice of the Andes
Casilla 17-17-691
Quito
ECUADOR

Radio Cairo
PO Box 1186
Cairo
EGYPT

Radio Estonia
21 Lomonossovi
200 100 Talinn
ESTONIA

Radio Finland
Box 10
SF-00241
Helsinki
FINLAND
or
Box 462
Windsor, CT 06095
USA
Phone (USA) 800-221-9539

Radio France International
B. P. 9516
F-75016 Paris Cedex 16
FRANCE

Deutsche Welle
PO Box 100 444
D-5000 Koln 1
GERMANY

Radio Netherlands
PO Box 222
1200JG Hilversum
HOLLAND

Radio Budapest
PO Box 1
H-1800
Budapest
HUNGARY

All India Radio
External Services Division
Parliament Street
PO Box 500
New Delhi 110 001
Delhi
INDIA

Voice of the Islamic Republic of Iran
IRIB External Services
PO Box 3333
Tehran
IRAN

Radio Iraq International
PO Box 8145
Baghdad
IRAQ

KOL Israel
PO Box 1082
91 010 Jerusalem
ISRAEL

RAI Radiotelevisione Italiana
Viale Mazzini 14
00195 Roma
ITALY

Radio Japan / NHK
2-2-1 Jinnan
Shibuya-ku
Tokyo 150-01
JAPAN

Radio Korea
Korean Broadcasting System
18, Yoido-dong
Youngdungpo-gu
Seoul
REPUBLIC OF KOREA

Radio Kuwait
PO Box 397
13004 Safat
KUWAIT

Radio Vilnius
Konarskio 49
LT-2674
Vilnius MTP
LITHUANIA

Voice of the Mediterranean
PO Box 143
Valletta
MALTA

Radio New Zealand International
Broadcast House
PO Box 2090
Wellington
NEW ZEALAND

Voice of Nigeria
P.M.B. 4003 Falomo
Ikoyi
NIGERIA

Radio Norway International
N-0340
Oslo 3
NORWAY

Far East Broadcasting Co.
Box 1
Valenzuela
Metro-Manila 1405
PHILIPPINES

Polish Radio Warsaw
External Services
PO Box 46, 00-950
Warsaw
POLAND

Radio Portugal International
Rua Sao Marcal 1
Lisbon
PORTUGAL

Radio Romania International
60-62 General Berthelot Street
PO Box 111
70749 Bucharest
ROMANIA

Radio Moscow
World Service
Moscow
RUSSIA

Broadcasting Service of the Kingdom of Saudi Arabia
PO Box 61718
Riyadh 11575
SAUDI ARABIA

Radio RSA
PO Box 91313
Auckland Park 2006
REPUBLIC OF SOUTH AFRICA

Spanish National Radio
Apartado 156.202
E-28080
Madrid
SPAIN

Radio Sweden
S-105 10 Stockholm
SWEDEN

Swiss Radio International
PO Box CH-3000
Berne 15
SWITZERLAND

Voice of Turkey
PK 333
06.443
Yenisehir
Ankara
TURKEY

Radio Ukraine International
Ul. Kreshchatik 26
252001 Kiev
UKRAINE

BBC World Service
PO Box 76
Bush House
London
UNITED KINGDOM

Voice of America
300 Independence Ave, S.W.
Washington, DC 20547
USA

Christian Science Monitor
Shortwave World Service
PO Box 860
Boston, MA 02123
USA

For The People (via WHRI)
Telford Hotel
3 River Street
White Springs, FL 32096
USA

KGEI—Voice of Friendship
1400 Radio Road
Redwood City, CA 94065
USA

KVOH—High Adventures Radio
PO Box 93937
Los Angeles, CA 90093
USA

WRMI-Radio Miami Internacional
PO Box 526852
Miami, FL 33152
USA

Radio Newyork International/Voyager Broadcast Services
c/o Allan H. Weiner
14 Prospect Drive
Yonkers, NY 10705
USA

WEWN-Eternal Word Network
PO Box 380247
Birmingham, AL 35238
USA

WHRI-World Harvest Radio
PO Box 12
South Bend, IN 46624
USA

WINB
PO Box 88
Red Lion, PA 17356
USA

WJCR
PO Box 91
Upton, KY 42784
USA

WRNO
Box 100
New Orleans, LA 70181
USA

WWCR-World Wide Christian Radio
1300 WWCR Avenue
Nashville, TN 37218
USA

WYFR-Family Radio
Family Stations, Inc.
290 Hegenberger Road
Oakland, CA 94621
USA

Vatican Radio
00120 Vatican City
VATICAN STATE

Appendix B
Radio listening clubs

Organizations of radio clubs

The following organizations can supply a detailed listing of radio clubs in the area they serve. Enclose return postage or International Reply Coupons with your letter to ensure a quick response.

ANARC
The Association of North American Radio Clubs
2216 Burkey Dr.
Wyomissing, PA 19610
USA

Association of Pan-Asian Radio Clubs
P.O. Chabdana-1702
Dt. Gazipur
BANGLADESH

EDXC
European DX Council
PO Box 4
St. Ives
Huntingdon, Cambs. PE17 4FE
ENGLAND

South Pacific Association of Radio Clubs
c/o NZ Radio DX League
212 Earn Street
Invercargill
NEW ZEALAND

National and international clubs

Joining a DX club is one of the best ways to keep up with current information on frequency changes, programming, etc. Most are operated by volunteers who have been enjoying their listening hobby for some time and would like to share their knowledge with others.

If you would like to receive a sample bulletin from any of these groups, send $2 US or 3 International Reply Coupons.

North American DX clubs

American Shortwave Listener's Club
Stewart MacKenzie, WDX6AA
16182 Ballad Lane
Huntington Beach, CA 92649
714-846-1685
shortwave

American Radio Relay League
225 Main Street
Newington, CT 06111
203-666-1541
amateur radio

* The Association of Clandestine Radio Enthusiasts (The A*C*E)
Kirk Baxter
PO Box 11201
Shawnee Mission, KS 66207
pirates and clandestines

* Association of DX Reporters
Reuben Dagold
7008 Plymoth Rd.
Baltimore, MD 21208
shortwave, ham, utility, AM, and longwave

Bearcat Radio Club
PO Box 291918
Kettering, OH 45429
1-800-423-1331
VHF/UHF scanning

* Canadian International DX Club
Sheldon Harvey, President
79 Kipps Street Road
Greenfield Park, Quebec T8G 1A5
CANADA
514-462-1459
all bands

* Longwave Club of America
45 Wildflower Road
Levittown, PA 19057
longwave - below 550 KHz

* Miami Valley DX Club
Box 291232
Columbus, OH 43229
shortwave

* National Radio Club
PO Box 5711
Topeka, KS 66605-0711
AM, medium wave

* North American Shortwave Association
Bob Brown, Executive Director
45 Wildflower Lane
Levittown, PA 19057
shortwave

* Radio Communications Monitoring Association
PO Box 542
Silverado, CA 92676
two-way communication, UHF/VHF scanning

* SPEEDX
(Society to Preserve the Engrossing Enjoyment of DXing)
Bob Thunberg, Business Manager
PO Box 196
DuBois, PA 15701-0196
shortwave broadcasts and utilities

* Worldwide TV-FM DX Association
PO Box 514
Buffalo, NY 14205
FM, VHF/UHF, satellites, and TV

Overseas DX clubs

African DX Association
c/o Friday Okoloise
Radio/Carrier Room
Nitel, Ashaka
Bauchi State
NIGERIA

ANDEX
Radio HCJB
PO Box 619
Quito
ECUADOR
shortwave

Australian DX Club
Box 227
Box Hill, Vic. 3128
AUSTRALIA
shortwave

British DX Club
54 Birkhall Rd.
Catford, London SE6 1TE
ENGLAND
shortwave

Czech DX Club
c/o Vasclav Dosoudil
Horni 9
CS-76821
Kvasice
CZECH REPUBLIC
shortwave

Danish Shortwave Clubs International
Taveager 31
DK-2670 Greve
DENMARK
shortwave

FRS Goes DX
PO Box 2727
6049 ZG Herton
THE NETHERLANDS
pirate radio activity in Europe and around the world

International Listeners Association
1 Jersey Str.
Hafod, Swansea SA1 2HF
ENGLAND
shortwave

International Shortwave League
6 Moorhead
Preston Upon The Weald Moors
Telford, Shropshire TF6 6DC
ENGLAND
shortwave and ham radio

Pakistan SW Listeners Clubs Association
Javaid Iqbal
PO Box 5
Sheikhupura 39359
PAKISTAN
shortwave

Pirate Chat
21 Green Park
Bath, Avon BA1 1HZ
ENGLAND
worldwide pirate activity on SW, AM, and FM

Pirate Connection
Kamnarsvagen 13D:220
22646 Lund
SWEDEN
European and North American pirate activity

Radio Budapest Shortwave Club
Radio Budapest
PO Box 1
H - 1800 Budapest
HUNGARY
quarterly paper mailed free to R. Budapest listeners who request it

South African DX Club
PO Box 72620
Lynwood Ridge
Transvaal 0040
SOUTH AFRICA
shortwave

Regional clubs

The following organizations are made up of people in the club's local area. Some focus on VHF/UHF /scanner/ frequency information, which would be of little or no use to people living elsewhere, and some clubs prefer to stay local so members can meet face-to-face to discuss their latest catches, show off QSL cards, equipment, and so on.

Most publish a bulletin or newsletter and will be happy to send you a sample copy for return postage.

All Ohio Scanner Club
Dave Marshall
50 Villa Road
Springfield, Ohio 45503-1036
VHF/UHF and ham

Association of Manitoba DXers
Shawn Axelrod
30 Becontree Bay
Winnipeg, Manitoba R2N 2X9
CANADA
204-253-8644
AM, shortwave, and VHF/UHF

Bay Area Scanner Enthusiasts
4718 Merdian Ave.
#265
San Jose, CA 95118
UHF/VHF scanning

Boston Area DXers
Paul Graveline
9 Stirling St.
Andover, MA 10810
508-470-1971
shortwave

Capitol Hill Monitors
Alan Henney
6912 Prince Georges Ave.
Takoma Park, MD 20912
UHF/VHF scanning

Chicago Area Radio Monitoring Association
Kurt Stoudt
2625 North Forest
Arlington Heights, IL 60004
UHF/VHF scanning

Chicago DX Club
c/o Thomas V. Ross
8225 West 43rd Place
Lyons, IL 60534
mostly shortwave

Cincinnati Area Monitoring Exchange
John Vodenik
513-398-5968
all bands

DX South Florida
3156 NW 39th St.
Ft. Lauderdale, FL 33309
shortwave

Frequency Fan Club
Race Scanning Monthly
PO Box 991
Mulberry, FL 33860
covers auto race scanner frequencies

Hawkeye Scanning Group of Iowa
PO Box 974-HS
Burlington, IA 52601-0974
VHF/UHF

Metro Radio System
Julian Olanskay
PO Box 26
Newton Highlands, MA 02161
617-969-3000
New England public safety frequencies

Michigan Area Radio Enthusiasts
Bob Walker
PO Box 311
Wixom, MI 48393
all bands

* Minnesota DX Club
PO Box 3164
Burnsville, MN 55337
all bands

Monitor Communications Group
Louis Campagna
8001 Castor Ave. #143
Philadelphia, PA 19152
all bands

Northeast Ohio DXers
Mike Fanderys
2802 North Ave.
Parma, OH 44134
216-661-2443
shortwave and utility stations

NE Ohio SW Listeners
Brian Schaft
317 South Rocky River Dr.
Berea, OH 44071
216-234-4628
shortwave

Northeast Scanner Club
Les Mattson
PO Box 62
Gibbstown, NJ 08027
UHF/VHF scanning - ME through VA

Ontario DX Association
Harold Sellers
PO Box 161 - Station A
Willowdale, Ontario M2N 5S8
CANADA
all bands

Radio Monitors Newsletter of Maryland
Ron Bruckman
PO Box 394
Hampstead, MD 21074
UHF/VHF scanning and shortwave

Regional Communications Network
Bill Morris
Box 83-M
Carlstadt, NJ 07072-0083
all bands - 50 mile radius of NY city

Rocky Mountain Radio Listeners
Wayne Heinen
4131 S. Andes Way
Aurora, CO 80013-3831
all bands - Denver metro area

Scanning Wisconsin
c/o AJC, Inc.
W. 17912 Pearl Drive
Muskego, WI 53150
VHF/UHF scanning

* Southern California Area DXers
Don R. Schmidt
3809 Rose Avenue
Long Beach, CA 90807-4334
310-424-4634
all bands

Toledo Area Radio Enthusiasts
Ernie Dellinger, N8PFA
6629 Sue Lane
Maumee, OH 43537
419-865-4284
all bands

Virginia Monitoring Digest
PO Box 34832
Richmond, VA 23234
all bands

* Washington Area DX Association
606 Forest Glen Road
Silver Springs, MD 20876
all bands

* ANARC members or associate members

Appendix C
Radio sources

These companies supply the radio receivers, transmitters, accessories, antennas, and books you'll need to get the most out of your listening hobby.

ACE Communications
Monitor Division
10707 E. 106th St.
Fishers, IN 46038
1-800-445-7717
VHF/UHF scanners

ACME Enterprises
1358 Coney Island Ave.
Suite 200 Z
Brooklyn, NY 11230
books on pirate radio

American Radio Relay League
225 Main Street
Newtingon, CT 06111
QST Magazine, ham radio books, ham license study guides

Antenna Supermarket
PO Box 563
Palatine, IL 60078
multi-band shortwave antennas

Antennas West
1500 North 150 West
Provo, UT 84605
1-800-926-7373
shortwave, ham, and scanner antennas

Austin Amateur Radio Supply
5325 North IH-35
Austin, TX 78723
ham radio transceivers, accessories, and repair service

Berry Electronics Supply Company
512 Broadway
New York, NY 10012
ham radio transceivers, accessories, and repair service

Chilton Pacific Ltd.
5632 Van Nuys Blvd. #222
Van Nuys, CA 91401
high-performance, long-range GE Superadio II (AM/FM portable)

Communications Electronics Inc.
PO Box 1045
Ann Arbor, MI 48106
ham, shortwave, and CB equipment and service

CQ Communications
76 N. Broadway
Hicksville, NY 11801
CQ ham radio magazine, CQ Amateur Radio Equipment Buyer's guide, CQ Antenna Buyer's Guide, ham radio books

CRB Research Books
PO Box 56
Commack, NY 11725
communications books, frequency guides for all bands

Drake Company, R. L.
PO Box 3006
Miamisburg, OH 45324
1-800-937-2534
shortwave radios

DX Radio Supply
PO Box 360
Wagontown, PA 19376
(215) 273-7823
National Scanning Report Magazine, Betty Bearcat Scanner guides, and other communications books

Electronic Equipment Bank
323 Mill Street, NW
Vienna, VA 22180
1-800-368-3270
ham radio transceivers, shortwave receivers, and repair service

Galaxy Electronics
67 Eber Ave.
Box 1201
Akron, OH 44309
216-376-2402
new and used shortwave radios and scanners

Gateway Electronics, Inc.
8123 Page Blvd.
St. Louis, MO
1-800-669-5810
books, and shortwave and ham radio kits

Gilfer Shortwave
52 Park Ave.
Park Ridge, NJ 07656
1-800-445-3371
shortwave radios and communications books

Grove Enterprises, Inc.
PO Box 98
300S. Highway 64 West
Brasstown, NC 28902
1-800-438-8155
Monitoring Times magazine, shortwave radios, scanners, books, and antennas

Gordon West Radio School
PO Box 2013
Lakewood, NJ 08701
ham radio license study guides, code practice cassettes, and classroom study supplies

Grundig
3520 Haven Ave.
Unit L
Redwood City, CA 94063
shortwave radios

Hamstuff
PO Box 14455
Scottsdale, AZ 85267
QSL storage boxes, ham radio T-shirts, etc.

Hamtronics, Inc.
4033 Brownsville Road
Trevosee, PA 19047
1-800-426-2820
ham radio equipment

HighText Publications, Inc.
125 N. Acacia Ave.
Suite 110
Solano Beach, CA 92075
books about shortwave listening, ham radio, and scanning

HR Bookstore
PO Box 209
Rindge, NH 03416
1-800-457-7373
ham and shortwave radio books

Hustler Antennas
One Newtronics Place
Mineral Wells, TX 76067
ham, scanner, and CB antennas

ICOM America, Inc.
2380 116th Ave, NE
Bellevue, WA 98004
ham radio transceivers and accessories

Index Publishing Group
3368 Governor Dr., Suite 273F
San Diego, CA 92122
1-800-546-6707
scanner, scanner modification, and secret radio frequency guidebooks Also sells books on electronic surveillance, hacking, and other aspects of the electronic underground

Kenwood USA Corporation
2201 E. Dominguez Street
PO Box 22745
Long Beach, CA 90801
shortwave receivers, ham radio transceivers, and accessories

Lentini Communications, Inc.
21 Garfield St.
Newington, CT 06111
1-800-666-0908
shortwave radios, scanners, and ham radio equipment

Marymac Industries
22511 Katy Fwy.
Katy (Houston) Texas 77450
1-800-231-2680
discount dealer of Radio Shack products

Midland International Corporation
1690 N. Topping Ave.
Kansas City, MO 64120
CB radios, VHF marine radios, antennas, and accessories

Mil-Spec Communications
PO Box 461
Wakefield, RI 02880
(401) 783-7106
shortwave radios and repair service

National Amateur Radio Association
PO Box 598
Redmond, WA 98073
1-800-468-2426
Amateur Radio Communicator magazine, license study materials

National Scanning Report
PO Box 291918
Kittering, OH 45249
1-800-423-1331
publishes National Scanning Report magazine

NuTechnology Newsletter/Hart Publishing
767 Xenon Court #117
Lakewood, CO 80228
publishes NuTechnology Newsletter and the Amateur Radio Mail Order Catalog and Resource Directory

Offshore Echos
PO Box 1514
London W72LL
ENGLAND
catalog of tapes, records, CDs, posters, etc. of offshore broadcast stations past and present, including, Radio Caroline and Radio New York International

QSLs by W4MPY
682 Mt. Pleasant Rd.
Monetta, SC 29105
QSL cards and logbooks

Palomar Engineers
PO Box 462222
Escondido, CA 92046
(619) 747-3343
antennas, pre-amplifiers, and filters

Philips
1-800-328-0795
call for dealer information on the Philips DC-777 AM/FM/Shortwave/Cassette car stereo

Popular Communications
76 North Broadway
Hicksville, NY 11801
Popular Communications magazine, Popular Communications' Communications Guide (annual)

Popular Electronics
PO Box 338
Mt. Morris, IL 61054
1-800-435-0715
Popular Electronics magazine

Radio Amateur Callbook
PO Box 2013
Lakewood, NJ 08701
ham radio callbooks, radio maps, license study materials

Radio Buffs
1-800-828-6433
call for price quotes on ham and shortwave books and equipment

Satman, Inc.
6310 N. University No. 3798
Peoria, IL 61612
1-800-472-8626
satellite TV equipment

Scanner World, USA
10 New Scotland Ave.
Albany, NY 12208
(518) 436-9606
scanners, shortwave radios, CBs, antennas, and radio books

Skyvision, Inc.
1050 Frontier Drive
Fergus Falls, MN 56537
1-800-334-6455
satellite TV receiving equipment

Snallygaster Press
PO Box 272
Springs, PA 15562
books on shortwave and pirate radio, also T-shirts from various pirates

Somerset Electronics, Inc.
1290 Hwy. A1A
Satellite Beach, FL 32937
1-800-678-7388
multi-mode decoders (for receiving digital transmissions from short-wave utility stations)

TAB/McGraw-Hill
Blue Ridge Summit, PA 17294-0214
1-800-822-8158
shortwave, ham radio, and antenna books

Tiare Publications
PO Box 493
Lake Geneva, WI 53147
specializes in books on pirate and clandestine radio

Tucker Electronics and Computers
PO Box 551419
Dallas, TX 75355-1419
1-800-527-4642
shortwave radios, ham radio equipment, computers, and test equipment

Turbo Electronics
PO Box 8034
Hicksville, NY 11834
1-800-33-TURBO
scanners, CBs, and other electronic equipment

Universal Radio
6830 Americana Pkwy
Reynoldsburg, OH 43068
1-800-431-3939
shortwave radios, decoders, ham radio, scanners, antennas, and books

W5YI Group
PO Box 565101
Dallas, TX 75356
1-800-669-9594
ham radio license study guides and code practice cassettes

Appendix D
Radio books

These books can help you get the most out of your radio listening hobby. Most are available from a number of electronics stores and shortwave mail-order companies.

Air Scan—Tom Kneitel

Frequency directory for aircraft monitoring. Includes frequencies on shortwave and scanner bands. (CRB)

ARRL Antenna Book

Information on nearly all types of antennas used by ham radio operators and shortwave listeners. 700 pages. (ARRL)

Discover DXing! An Introduction to TV FM and AM DXing—John Zondlo

Helps the beginners discover how to tune in distant stations on the AM, FM, and TV bands. Contains info on equipment, propagation, antennas, and a section of "best bets" for hearing all 50 states on the AM band. (Universal Radio Research)

Easy-Up Antennas For Radio Listeners and Hams—Edward Noll

Tells how to construct low-cost, easy-to-erect antennas for LW, MW, FM, SW, ham, and scanner frequencies. Explains the latest antenna designs and construction tips, tools, and techniques. 157 pages. (Sams)

Guide to Utility Stations—J. Klingenfuss

Annual reference book lists frequencies of utility stations around the world—3,590 call signs and 19,549 frequencies. Covers utility stations that use AM, SSB, CW, RTTY, and new modes. 537 pages. (Klingenfuss)

Guide to World RTTY Stations—J. Klingenfus

Comprehensive list of radioteletype stations from 3 to 30 MHz, including military, press, air, meterological, maritime, and diplomatic stations. 105 pages. (Klingenfuss)

HF Aeronautical Communications Handbook—R. E. Evans

The standard reference book on hearing aeronautical communications on the shortwave frequency bands. 226 pages. (Universal Radio Research)

International Callsign Handbook—Gayle Van Horn

List of U.S. and foreign call signs and their identifications. Includes military, NASA, customs, etc. Also lists longwave CW beacons. 248 pages. (Grove)

National Radio Club AM Radio Log—NRC

Accurate and comprehensive guide to AM stations in the U.S.A. and Canada. Includes addresses for all stations. 350 pages. (National Radio Club)

Now You're Talking!—American Radio Relay League

Everything you need to know to get your code-free Technician and Novice-class licenses. (ARRL)

Passport to World Band Radio

This annual book features schedules, station addresses, receiver reviews, and information on shortwave listening in general. About 390 pages. (IBS)

Pirate Radio Directory—George Zeller

An annual yearbook of the North American pirate radio scene. Includes illustrations of pirate QSLs and information about the broadcasting activity of nearly 100 pirate stations. 80 pages. (Tiare)

Pirate Radio Stations: Tuning in to Underground Broadcasts—Andrew R. Yoder

A behind-the-scenes look at pirate radio in North America and Europe. Explains what pirates are, how to hear them, and how to obtain QSL cards from pirate radio stations. 177 pages. (TAB/McGraw-Hill)

Scanner and Shortwave Answer Book—Bob Grove

Bob Grove answers questions concerning the listening and technical aspects of scanner and shortwave listening. 152 pages. (Grove)

Scanner Radio Guide—Larry M. Barker

Explains what scanners are, what they're capable of receiving, and the "skip" conditions that sometimes make it possible to hear transmissions from hundreds of miles away. Contains several useful frequency charts. 150 pages. (HighText)

Secrets of Successful QSLing—Gerry Dexter

Explains how to improve your QSL return rate. 120 pages. (Tiare)

Shortwave Listening Guidebook—Harry Helms

Tells you what shortwave radio is, how to select and use one, what you'll hear on the shortwave bands, how to obtain program schedules, etc. 320 pages. (HighText)

Shortwave Propagation Handbook—Jacobs and Cohen

Explains how shortwave propagation works: ionosphere, sunspot cycles, propagation forecasting, and unusual HF and VHF propagation conditions. 150 pages. (CQ Communications)

Shortwave Radio Listening for Beginners—Anita Louise McCormick

Explains everything the new shortwave listener needs to know to tune in the world. Also introduces you to long-distance AM reception, ham radio, and scanning. 175 pages. (TAB/McGraw-Hill)

Shortwave Receivers Past and Present—Fred Osterman

Reviews over 200 shortwave receivers that were manufactured in the past 20 years. 106 pages. (Universal Radio)

The Aeronautical Communications Handbook—R. E. Evans

Reference book on aeronautical communications on the shortwave bands. 266 pages. (Universal Radio)

The Complete Shortwave Listener's Handbook—Bennett, Hardy, and Yoder

Gives shortwave hobbyists all the info they need to tune in stations from around the world. Topics include receivers, antennas, propagation, QSLing, shortwave stations, etc. 350 pages. (TAB/McGraw-Hill)

The Radio Hobbyist's Buyer's Bluebook—Bob Grove

Lists new and typical used prices for hundreds of current and discontinued shortwave radios and scanners. (Grove)

Scanners and Secret Frequencies—Henry Eisenson

Scanner guide for hobbyists of all levels. Reviews many models of scanners and lists hundreds of frequencies for services, such as law enforcement, aviation, fire, cordless and cellular phones, baby monitors, government, and industry. 320 pages. (Index Publishing Group)

The SWL's Antenna Handbook—Robert J. Traister

Complete guide to understanding and building shortwave antennas. Includes ten antenna construction projects. 191 pages. (TAB/McGraw-Hill)

The RTTY Listener—Fred Osterman

Information about advanced RTTY and FAX monitoring. 222 pages. (Universal Radio)

The Traveler's Guide to World Radio

Annual pocket-size guide to English language broadcasts to 50 of the world's most popular destinations. 128 pages. (Billboard Publications)

Tune In On Telephone Calls—Tom T. Kneitel

Explains how telephone calls on cordless phones, car phones, ship phones, etc. can be heard on scanners and shortwave radios. 160 pages. (CRB)

World Press Service Frequencies—Thomas Harrington

Lists RTTY press stations by time, frequency, country, and press service. 84 pages. (Universal Electronics)

World Radio TV Handbook

Annual book of shortwave, medium wave (AM), and TV station schedules, addresses, etc. Contains radio reviews and information on shortwave listening, DX clubs, and more. Approximately 608 pages. (Billboard)

Worldwide Aeronautical Communications—R. E. Evans

Covers aeronautical communications on all major civil and military frequencies from 2 to 26 MHz, cross-referenced by country. 42 pages. (Universal Radio Research)

Glossary

aeronautical radio Communications between aircraft and ground control, or between one aircraft and another.

amateur call letters Ham radio identification letters. In the USA, they are issued by the Federal Communications Commission.

antenna Wire or other metal device used to gather radio waves and direct them into a receiver.

ARRL (American Radio Relay League) The largest ham radio organization in the United States. Publishes QST magazine.

ASCII (American Standard Code for Information Exchange) Most commonly used code for exchanging information via computer.

bandwidth Amount of frequency space taken up by a signal.

beam antenna A directional antenna used by many hams.

BFO (Beat Frequency Oscillator) A feature on some receivers that makes single side band voice transmissions intelligible, and improves Morse Code reception.

clandestine station Station broadcasting political messages without a license, usually in connection with revolutionary movements.

digital readout Display on a radio receiver that shows exactly which frequency you are on.

domestic station Station that broadcasts programs intended for a local or regional audience.

downlink Frequency used to relay satellite transmissions to earth.

DX Distance.

DXer Person who enjoys listening to distant radio transmissions.

DX club Organization for people interested in hearing distant radio transmissions. Many DX clubs publish monthly bulletins for their members.

DX program Program on a shortwave station that provides listeners with up-to-the-minute information on international broadcasts.

electromagnetic waves Energy waves made up of an electrical and magnetic field, used to transmit radio and TV signals.
feed horn Part of a satellite dish antenna that collects the reflected signal and funnels it into the amplifier.
footprint Area covered by satellite transmissions.
free radio station Pirate (unlicensed) radio station.
frequency Number of radio waves that pass a given point per second.
geomagnetic storms Disruption of earth's magnetosphere caused by solar flares.
geostationary orbit An orbit 22,300 miles above us where satellites circle our planet at the same speed as the Earth's rotation.
GHz (Gigahertz) Billion hertz, or radio waves, per second.
GMT (Greenwich Mean Time) Old term for UTC, the world-wide standard time zone used in shortwave radio.
ground wave Radio wave that stays near the earth, and can be heard only for a limited distance.
HF (high frequency) 3 to 30 MHz (includes shortwave bands).
IRC (international reply coupon) World-wide exchange medium used to pay for postage costs.
ionosphere Layers of electrically charged gas in our atmosphere that affect radio wave skip.
ITU (International Telecommunications Union) International organization that regulates use of the radio wave spectrum.
ITU Phonetics World-wide system of phonetics, recommended by the ITU for use in voice communications when reception conditions are difficult.
kHz (kilohertz) Thousand hertz, or radio waves, per second.
line of sight Straight, non-curving path radio waves take on VHF and UHF frequencies.
LSB (Lower Side Band) Single sideband transmission with the upper sideband and carrier removed.
LUF (Lowest Useable Frequency) The lowest frequency that can be used to transmit from one location to another.
magnetosphere Magnetic field that surrounds the earth and protects us from much of the sun's harmful radiation.
medium wave (mw) Frequencies used for AM band broadcasts.
MHz (Megahertz) Million hertz, or radio waves, per second.
mode Method of transmitting, such as AM, FM, Morse Code, and radioteletype.
MUF (Maximum Useable Frequency) The highest frequency you can use to transmit from one area to another.
multi-mode decoder Device used to interpret digital transmissions (CW, RTTY, ASCII, and other modes) so you can view them on a digital display, television, computer screen, or print-out.
offshore station Radio station that operates from a ship—usually in international waters.
OWF (Optimum Working Frequency) Frequency that can give the best signal from a given area, to a given area.
pirate station Hobby broadcast station, operated without a license.

propagation Transportation of radio waves through the atmosphere from one part of the world to another.

QRP Low-power operation (in ham radio, usually under 10 watts).

Q signals Ham radio abbreviations for frequently used messages.

QSL card A card from a radio station that verifies you heard their broadcast.

reception report Report of signal quality sent to a station, usually for the purpose of obtaining a QSL card.

relay station A station that picks a signal up on one frequency (usually from a satellite downlink) and retransmits it on another (AM, FM, or shortwave).

repeater Automated system used to pick up, amplify, and re-transmit a signal.

RST (readability, signal, tone) Signal reporting system used by ham radio operators.

RTTY (Radioteletype) Mode of digital transmission used by hams and utility stations.

scanner A radio that scans through pre-programmed frequencies (usually in the VHF-UHF range) to locate transmissions of police calls, fire calls, etc. Some shortwave radios also have scanning capability.

selectivity The ability of a receiver to separate one station from another nearby broadcaster.

sensitivity The ability of a receiver to amplify weak signals so you can better hear them.

shortwave (sw) Radio frequencies between 3 MHz and 30 MHz, also HF.

SINPO (signal, interference, noise, propagation, overall) Signal reporting system used by shortwave listeners.

skip zone The area where you are too far away from the station to hear the ground wave, but too close to receive the skip.

solar flare A powerful explosion on the sun that can cause blackouts in shortwave radio reception here on Earth.

solar flux index A method of measuring the solar activity that influences radio wave propagation.

SSB (single side band) Mode of transmission used by ham radio operators, utility stations and a few shortwave stations. One sideband is cut off to save on bandspace and must be replaced by using the SSB/BFO control on your receiver.

UHF (Ultra High Frequency) 300 MHz to 3 GHz.

Uplink Frequency used to transmit signals from an earth station up to a satellite.

USB (Upper Side Band) Single side band transmission with the lower side band removed.

utility stations A wide assortment of government and commercial stations that don't broadcast to the general public, but operate in order to transmit, to convey, or exchange information with each other. Aviation, ship-to-shore, international weather, and military transmissions are included in this category.

VHF (Very High Frequency) 30 MHz to 300 MHz.

wavelength The distance from the peak of one radio wave to the peak of the next.

WPM (Words Per Minute) Number of words sent per minute (Morse Code).

XCVR (transceiver) Receiver and transmitter in one unit.

Index

D

E

F

G

H

I

T

U

V

About the Author

Anita Louise McCormick has more than 20 years' experience as a shortwave radio listener. She is a ham radio operator (KA8KGI) and freelance writer who has written articles on shortwave radio listening for such magazines as *Listen*, *Mature Living*, and *Pioneer*. She is also the author of *Shortwave Radio Listening for Beginners*, published by TAB/McGraw-Hill.

DATE			